Albert

MW00878679

Father of the Modern Scientific Age

By Michael W. Simmons

Table of Contents

Chapter One: Einstein's Origins

"It is quite clear to me that the religious paradise of youth, which was thus lost, was a first attempt to free myself from the chains of the "merely-personal," from an existence which is dominated by wishes, hopes, and primitive feelings. Out yonder there was this huge world, which exists independently of us human beings and which stands before us like a great, eternal riddle, at least partially accessible to our inspection and thinking. The contemplation of this world beckoned like a liberation, and I soon noticed that many a man whom I had learned to esteem and to admire had found inner freedom and security in devoted occupation with it."

Albert Einstein, "Autobiographical Notes"

Family

Until the 1870's, Germany, as a nation, did not exist. The German-speaking regions of western Europe constituted a loose alliance of small kingdoms, duchies, princedoms, and free cities, with no central leaderships. Chancellor Otto von Bismarck united the various German regions, apart from Austria, under the Prussian king, Wilhelm I, in a war against France, and this political alliance of territories has survived into the modern age as the German state.

One of these united German territories was Swabia, a region which to this day preserves a distinct linguistic and ethnic identity within the larger Germany. In the nineteenth century, Swabia was so well known for being an area of Jewish settlement that, in later decades, Nazi propaganda films would single the region out as an example of German territory that had been overrun by Jews.

One Jewish family, which by the time of Germany's unification had been settled in Swabia for several generations, was the Einsteins. Hermann Einstein was born in 1847 in the town of Buchau, where the Jewish community was particularly strong. (The legal status of Jews in Europe was unstable throughout the nineteenth century, so the existence of a thriving Jewish settlement was a blessing not taken for granted by the Jews who lived there.) Hermann Einstein was born just after Jews in his region began to be permitted to practice skilled trades, but long before Jews were permitted to attend most European universities. Like his famous son, Hermann displayed a marked facility for mathematics as a boy, but as there were no opportunities for advanced study in Buchau, his family arranged for him to attend high school in Stuttgart, some seventy five miles away. This was the end of his formal education however, as his family did not have the means to pay for university, even if one were willing to admit him.

As a young man, Hermann moved with his aging parents from remote Buchau to the more industrial city of Ulm, where his cousin owned a business selling feather beds. Hermann Einstein was by all accounts a gentle and docile person, clever, but inclined to passivity; possibly this is why he made a dreadful salesman, with a poor head for financial matters. In 1876, at the age of twenty nine, he married the eighteen year old Pauline Koch, whose father was a prosperous grain merchant. Pauline was considerably more vibrant, practical, and forthright than her husband, but the marriage appears to have been a very happy one.

On Friday, March 24, 1879, in Ulm, Pauline and Hermann Einstein had their first child, a son. Initially, they had wanted to name him Abraham, for Hermann's father, but they decided that the name would mark him too obviously as Jewish. They called him Albert instead; if he could not share his grandfather's name, at least he could share his first initial.

Hermann Einstein had a large number of siblings, and shortly after Albert's birth, when Hermann began suffering serious business setbacks, one of them came to the family's rescue: Albert's uncle Jakob invited the family to join him in Munich, where Hermann could make a living working for his brother's gas and electrical company. Jakob was the one member of the family who had managed to gain admittance to a university, or rather a technical school, and he had earned an engineering degree. Jakob's scientific training enabled him to oversee the technical aspects of the business, while Hermann was in charge of getting contracts from the German government and various business to supply gas and electrical power.

Developmental Years

Some sources on Einstein's early life claim that Hermann and Pauline were concerned about their son's development when he was a small child. Many biographies of Einstein's life claim that, for the first two years of his life, an age during which most children are at least beginning to babble and build a small vocabulary, he spoke not a single word. According to those sources, this led his parents to fear that he was not "normal". However, in 1881, his younger sister and lifelong best friend Maja (a nickname for "Maria") was born, and her birth supposedly occasioned the utterance of Einstein's first spoken sentence. Apparently, the three year old boy had received the impression that the birth of a younger sibling—a person with

whom he was repeatedly told he could "play"—
meant that he was receiving a new toy. On being
presented with a baby girl, he shocked his
parents by inquiring, "But where are the
wheels?" However, he presumably came to
understand that eventually his younger sister
would be able to move herself from place to
place—no wheels required.

There is some room to speculate that this story
was merely an amusing anecdote invented by
Einstein's family to give color to his childhood.
Though Einstein was a prolific writer, he told few
stories about his earliest years. Despite the fact
that he had a prodigious memory for everything
related to his scientific work, he seems not to
have retained detailed memories about his
childhood, which has opened the door to
fabrication and speculation by the many
biographers who have naturally been eager to
understand even the smallest details of the
intellectual development of one of the most

brilliant scientists in history. And even within Einstein's own family, there are conflicting legends about his childhood language development. Einstein himself, for example, claimed, somewhat humorously, that the reason he did not speak until his third year was that he chose not to—that he found the stilted baby talk expected of him at that age as intellectually beneath. But his grandmother claimed that she had many conversations with him as two year old in which they discussed all of his childish ideas and theories about the world. The exact truth is probably beyond the recall of history at this point.

One of the most famous speculations about Einstein in recent decades, owing largely to the anecdotes about his language delays, is that he fell somewhere on the autism spectrum. Autism is a neurological condition that is sometimes diagnosed in children who experience language delays, have difficulty regulating sensory input, and appear uninterested in forming social relationships with other children their age. It does seem to be true that as a child Einstein at least suffered from the self-imposed isolation and loneliness of an extremely bright child who lacked intellectual equals. He displayed little interest in joining in with the games of the other children in his middle class Munich suburb. So withdrawn was he, and so wrapped up in private puzzle games, that his childhood governess nicknamed him "Father Bore", as if his detachment made him seem much older than he actually was. In later years, one of Einstein's colleagues remarked that "From the very beginning he was inclined to separate himself from children his own age and to engage in daydreaming and meditative musing."

Einstein was also subject to violent fits of temper as a young child, in which his face turned strange colors, and his rage provoked him to throw hard objects at his tutor and even his younger sister.

Fortunately, these violent outbursts did not follow him into adulthood. Autistic children do sometimes display uncontrolled outbursts, often provoked by excessive sensory stimulation. But retrospectively assigning medical, psychological, and neurological diagnoses to famous people in history is a notoriously imprecise business.

Some famous autistic writers, such as Temple Grandin, have claimed Einstein as an important model for the autistic community, an example of how brilliance is often tied to the eccentricities or odd behaviors that are often stigmatized in autistic people. But other autistic writers have questioned the anecdotal evidence that forms the basis on Grandin's diagnosis, such as Einstein's proclivity for wearing green slippers to his job at the patent clerk's office where he so famously wrote his first papers on relativity. There is a strong argument that his two marriages, his many friendships with other scientists, and his tendency toward introversion constitute evidence against an autism diagnosis. Again, there is probably no way to be certain at this late date in history. Autism was not theorized until 1908 and was not formally studied until 1934, so no diagnostic model existed to evaluate Einstein in his childhood.

It may be worth noting, however, that after Einstein's death, his brain was removed from his body by pathologist Thomas Harvey, of Princeton University, so that it could be stored for scientific analysis. It was dissected and preserved in two hundred and forty pieces between sheets of a material called collodion, which is similar to plastic. Einstein's brain has been widely photographed and studied, and has been subjected to magnetic resonance imaging in recent years. There was, naturally, a great deal of speculation that the brain of such a brilliant person might possess detectable abnormalities. The question of whether Einstein's brain does, in fact, display these abnormalities is controversial, but some scientists have claimed that his parietal lobes were at least fifteen percent larger than is found in most people. By contrasts, the parietal lobes of individuals with autism are often found to be underdeveloped, which would seem to be evidence that contradicts a retrospective diagnosis of autism in Einstein.

Though Einstein wrote little about his early life in purely autobiographical terms, despite being asked to do so repeatedly, in his sixties he wrote an article for the Saturday Review in which he described a few incidents in his childhood which not only stand as examples of the power of his intellect at that age, but which penetrate even further by attempting to account for the way Einstein thought, the way in which his imagination influenced his understanding of the world throughout his life. In this article, he recounts how, as a child, lying ill in bed, his father came to visit him, and to distract him from feeling unwell, showed him a compass. The behavior of the compass presented such a powerful mystery to Einstein that he later said that his whole body shook from the shock of it. As he tells it,

"A wonder of such nature I experienced as a child of four or five years, when my father showed me a compass. That this needle behaved in such a determined way did not at all fit into the nature of events which could find a place in the unconscious world of concepts (effect connected with direct "touch"). I can still remember — or at least believe I can remember — that this experience made a deep and lasting impression upon me. Something deeply hidden had to be behind things. What man sees before him from infancy causes no reaction of this kind; he is not surprised over the falling of bodies, concerning wind and rain, nor concerning the differences between living and non-living matter."

When Einstein speaks of the "wonder" he felt when he looked at the compass needle, he means that it had never before occurred to him that there was more to the universe than he could see and experience with his own senses. But when the needle moved, seeking magnetic north, he intuited, in a way, the presence of magnetic fields—or at least, of *some* force that made objects in his universe behave in a specific, predictable manner. He had, by that age, already dispensed with the idea of invisible forces such as God—his family, though Jewish, was no religiously observant, and as a child Einstein determined quickly that the stories he heard from the Bible could not possibly be true. But the presence of natural physical forces acting on the world around him was evidence by the sight before his eyes, and the wonder he experienced at that moment would set the course for his life's work.

In the same article, Einstein relates another formative experience from his childhood that created in him a sense of wonder, this time pertaining to mathematics.

"At the age of twelve I experienced a second wonder of a totally different nature: in a little book dealing with Euclidian plane geometry, which came into my hands at the beginning of a school year. Here were assertions, as for example the intersection of the three altitudes of a triangle in one point, which — though by no means evident — could nevertheless be proved with such certainty that any doubt appeared to be out of the question. This lucidity and certainty made an indescribable impression upon me. That the axiom had to be accepted unproved did not disturb me. In any case it was quite sufficient for me if I could peg proofs upon propositions the validity of which did not seem to me to be dubious.

"For example, I remember that an uncle told me the Pythagorean theorem before the holy geometry booklet had come into my hands. After much effort I succeeded in "proving" this theorem on the basis of the similarity of triangles; in doing so it seemed to me "evident" that the relations of the sides of the right-angled triangles would have to be completely determined by one of the acute angles. Only something which did not in similar fashion seem to be "evident" appeared to me to be in need of any proof at all. Also, the objects with which geometry deals seemed to be of no different type than the objects of sensory perception, "which can be seen and touched." This primitive idea, which probably also lies at the bottom of the well-known Kantian problematic concerning the possibility of "synthetic judgments a priori" rests obviously upon the fact that the relation of geometrical concepts to objects of direct experience (rigid rod, finite interval, etc.) was unconsciously present.

"If thus it appeared that it was possible to get certain knowledge of the objects of experience by means of pure thinking, this "wonder" rested upon an error. Nevertheless, for anyone who experiences it for the first time, it is marvelous enough that man is capable at all of reaching such a degree of certainty and purity in pure thinking as the Greeks showed us for the first time to be possible in geometry."

Like some other extraordinary scientists who made an effort to explain to the world what the universe looked like through their eyes, Einstein describes a mental process that challenges the imaginations of less gifted people. An inventor and contemporary of Einstein's, Nikola Tesla, wrote in his autobiography that, from infancy, he had visions, of the literal rather than the mystical variety. He could visualize anything he had seen just as clearly as if the object or person he was imagining stood before him. Later, he learned to visualize things he had made up in his head in equally profound detail; when he began inventing, he found that he did not need draft his machines or even build prototypes to know whether or not they would work, because he could see them in his mind so clear.

Einstein was a physicist, not an inventor, and his powers of imagination seem not to have had quite so practical a bent, but he too thought in terms of pictures and images. In fact, he was dubious as to whether the process of seeing the world in terms of imagines could even be called thinking, which he deemed a word-oriented process. As he describes it,

"In a man of my type the turning-point of the development lies in the fact that gradually the major interest disengages itself to a far-reaching degree from the momentary and the merely personal and turns towards the striving for a mental grasp of things.

"What, precisely, is "thinking"? When, at the reception of sense-impressions, memory-pictures emerge, this is not yet "thinking." And when such pictures form series, each member of which calls forth another, this, too, is not yet "thinking." When, however, a certain picture turns up in many such series, then — precisely through such return — it becomes an ordering element for such series, in that it connects series which in themselves are unconnected. Such an element becomes an instrument, a concept- I think that the transition from free association or "dreaming" to thinking is characterized by the more or less dominating role which the 'concept" plays in it. It is by no means necessary that a concept must be connected with a sensorily cognizable and reproducible sign (word); but when this is the case thinking becomes by means of that fact communicable...

"For me it is not dubious that our thinking goes on for the most part without use of signs (words) and beyond that to a considerable degree unconsciously. For how, otherwise, should it happen that sometimes we "wonder" quite spontaneously about some experience? This "wondering" seems to occur when an experience comes into conflict with a world of concepts which is already sufficiently fixed in us. Whenever such a conflict is experienced hard and intensively it reacts back upon our thought world in a decisive way. The development of this thought world is in a certain sense a continuous flight from "wonder."

Einstein's childhood was not entirely devoted to his budding scientific interests, however. His mother, Pauline, insisted that he study music.

She was herself a talented pianist, and she sensed that her gifted son possessed the tenacity and discipline necessary to make an excellent musician. She arranged for him to take violin lessons, and though he was initially rebellious, he formed a deep appreciation for the sonatas of Mozart that stayed with him the rest of his life. Though Einstein only took lessons between the ages of six and thirteen, he never stopped playing, and he remarked once that if he had not been destined for a career in the sciences, he would likely have been a professional musician. "Life without playing music is inconceivable for me," he explained. I live my daydreams in music. I see my life in terms of music…I get most joy in life out of music."

According to his second wife, Elsa, music also played a direct role in Einstein's scientific work. "As a little girl, I fell in love with Albert because he played Mozart so beautifully on the violin," she is known to have remarked. "He also plays the piano. Music helps him when he is thinking about his theories. He goes to his study, comes back, strikes a few chords on the piano, jots something down, returns to his study." Einstein played to such a high standard that he was often asked to give performances at benefit concerts for charitable causes. After one such concert, a music critic who knew that Einstein was world famous, but did not know why, remarked though Einstein played excellently, he did not deserve the level of fame he enjoyed, there were many other violinists who played just as well. He had no idea that Einstein was a physicist, and that music was only a hobby for him.

Einstein and Judaism

Because there were no Jewish schools near his home, and his family was not religious in any case (Hermann Einstein was prone to referring to Jewish religious rituals as "ancient superstitions") Einstein attended Catholic school as a boy. Ironically, he excelled at his compulsory religious studies class. As an adult, he recalled that the teachers were kind and not inclined to be prejudiced towards their only Jewish student, but that anti-Semitic prejudice was strong amongst the other children. He was bullied, attacked by gangs of students on his walks home, and at the age of nine he was moved to a high school with an advanced curriculum, where math and science, as well as Latin and Greek, were emphasized in the curriculum. There, for the first time, he was given religious instruction in Judaism by a teacher who had been specially hired for the benefit of Einstein and a few other Jewish students.

The ostracism he had experienced at his first school made Einstein feel like an outsider, and it is perhaps because of this that, once he was given the opportunity to make a study of Judaism, Einstein became enthusiastically observant of

Jewish religious practices. His sister claimed that he began to eat kosher and observe the Sabbath, which was somewhat difficult for him as no one else in his family was doing so. Perhaps this was an early attempt on Einstein's part to explain to himself the deep mysteries of the universe, such as he had observed in watching the compass needle spin.

In any case, his enthusiasm for religious life only lasted until the age of twelve. Around the time when he would have begun preparing for his bar mitzvah, the Jewish coming of age ceremony that marks the passage from childhood to young adulthood, Einstein came under the competing, and eventually eclipsing influence of scientific study.

According to Einstein biographer Walter Isaacson, the Einstein family adhered to at least one Jewish custom: prosperous families would invite "needy religious scholars" to eat dinner with them on the Sabbath. Reflecting their secularism, the Einsteins opted instead to invite a Jewish student from a nearby medical school to eat dinner with them on Thursdays. The name of this student Max Talmud (later Anglicized to Talmey); he was twenty one when he first met the ten year old Einstein, he played a profound role in directing Einstein's attention to scientific studies. Einstein borrowed a twenty one volume series on natural science from Talmud, and his interest in religion began to wane.

"Through the reading of popular scientific books, I soon reached the conviction that much in the stories of the Bible could not be true," Einstein remarked of this period in his life. "The consequence was a positively fanatic orgy of freethinking, coupled with the impression that youth is intentionally being deceived by the state through lies."

Though he never again considered himself religious, the political climate of Europe during Einstein's lifetime naturally forced him into a complicated relationship with his Jewish identity. The new nation of Germany, united by Bismarck under the rule of Kaiser Wilhelm I, was military mad. There had been war with France shortly before Einstein's birth, a war that had been deliberately provoked by Bismarck so that he could convince the German states to rally under the banner of German identity. Even after the war, Germany only increased its militarism, determined to make itself impregnable to attack. (This military build-up in the late nineteenth century led directly to the outbreak of World War I, as Wilhelm II came to the throne, and Bismarck's careful balance of power diplomacy unraveled under his aggressive leadership.)

Einstein grew deeply disenchanted with Germany and his German identity during the years leading to World War I. His family would move to Switzerland when Einstein was seventeen, and from that point forward he ceased to identify as German. Even after Einstein began to grow famous for his special theory of relativity, which he published in 1905, he was the victim of anti-Semitism, as when he applied for a post at the University of Prague; he believed that he was denied because of being Jewish. Despite his own relative indifference towards Jewish religion, he was forced to dwell deeply on the implications of his Jewish identity because of the treatment he received. In 1914, Einstein moved to Berlin, and his relativity theory swiftly fell under attack, primarily because its author was Jewish. In the late nineteen twenties and nineteen thirties, he was targeted by the Nazi party, and eventually left Berlin after he was warned that the National Socialist party was calling for his assassination.

During the second World War and afterwards, Einstein was approached by Zionist

organizations about lending his support to the
formation of the new Jewish state in Palestine.
Einstein had deeply mixed feelings about this.
After his experiences in pre-war Germany, he
had a deep loathing and antipathy towards
nationalism of any kind. Even after the
Holocaust, he was supportive of the rights of
Palestinian Arabs, and was concerned that it
would be impossible for Israeli Jews to exist
alongside them peacefully. He understood,
however, the longing for a homeland where Jews
would be free of persecution. He was
instrumental in raising funds for the
establishment of The Hebrew University in 1918,
and at one point, owing to his status as "a living
patron saint for Jews" (according to biographer
Walter Isaacson) he was offered the presidency
of Israel, which he declined.

Despite his refusal to identify himself with a
religion, Einstein retained a belief in God,

though not an orthodox one. In 1930, Einstein wrote,

"The most beautiful emotion we can experience is the mysterious. It is the fundamental emotion that stands at the cradle of all true art and science. He to whom this emotion is a stranger, who can no longer wonder and stand rapt in awe, is as good as dead, a snuffed-out candle. To sense that behind anything that can be experienced there is something that our minds cannot grasp, whose beauty and sublimity reaches us only indirectly: this is religiousness. In this sense, and in this sense only, I am a devoutly religious man."

Ultimately, Einstein's Jewish identity was tied neither to religion nor Zionism, but it remained an inextricable and paramount aspect of his identity. As journalist Paul Berger puts it,

"[...] Einstein felt a deep bond with the Jewish people, one that transcended religion as well as the anti-Semitism that Nazis and their sympathizers leveled against him.

"Despite his differences with fellow Jews in their practices and beliefs, Gimbel says, Einstein recognized that the relationship he had with Jewish friends was fundamentally different from the one he had with non-Jewish friends. He came to believe that whatever their beliefs, Jews shared common traits.

"The first trait Einstein identified as common among Jews was an ability to face the world with a sense of awe and joy, whether the Jews in question were rapturous Hasidim or secular physicists.

"The second trait Einstein identified was a sense of social justice. As he wrote in 1938, "The bond that has united the Jews for thousands of years and that unites them today is, above all, the democratic ideal of social justice coupled with the ideal of mutual aid and tolerance among all men."

Education

One of the most popular myths about Albert Einstein is that he failed algebra as a twelve year old. It is unclear where that myth originated, but it has been so widely repeated as to be received as unquestioned fact in all manner of popular writing on the subject of Einstein's education.

The truth is far less ironic: Einstein excelled in math and, as he told an amused friend in

Princeton who showed him a copy of a newspaper article with the headline "Greatest Living Mathematician Failed in Mathematics", he had mastered differential and integral calculus before he was fifteen. At school, his abilities in mathematics were always rated as far beyond what was expected of students his age. Throughout his years in lower schools, he was always ranked first in his class. As a young child, Einstein taught himself advanced forms of math from books that his parents purchased for him as gifts, and he studied alone throughout his summer vacations rather than play with other children.

His uncle Jakob, the engineer who had invited Einstein's father to join him in Munich when Hermann's feather bed company failed, was instrumental in whetting his nephew's appetite for advanced mathematics. He employed a striking metaphor when he introduced Einstein to algebra, calling it "a merry science": "When the animal we are hunting cannot be caught, we call it X temporarily and continue to hunt until it is bagged." It was also Jakob who first introduced Einstein to the Pythagorean theorem, which so arrested his attention and captured his imagination.

Max Talmud, the poor medical student who presented the ten year old Einstein with the *People's Books on Natural Science* series of books by Aaron Bernstein, was responsible for first cultivating Einstein's interest in physics. According to biographer Walter Isaacson, he was captivated by Bernstein's thoughts on the speed of light:

"[...] Bernstein asked readers to imagine being on a speeding train. If a bullet is shot through the window, it would seem that it was shot at an angle, because the train would have moved between the time the bullet entered one window and exited the window on the other side. Likewise, because of the speed of the earth through space, the same must be true of light going through a telescope. What was amazing, said Bernstein, was that experiments showed the same effect no matter how fast the source of the light was moving. In a sentence that, because of its relation to what Einstein would later famously conclude, seems to have made an impression, Bernstein declared, "Since each kind of light proves to be of exactly the same speed, the law of the speed of light can well be called the most general of all nature's laws."

Max Talmud assisted Einstein in his independent studies for as long as he was able, but there came a point where the child surpassed the learning of the university student, as least in mathematics. Talmud began tutoring Einstein in philosophy after that, encouraging him to read the works of German philosopher Immanuel Kant.

Despite flourishing in his extracurricular studies, Einstein was growing deeply unhappy at his Munich high school. The militarism of Bismarckian German society depressed him; an explicit aspect of the teaching at his school was unquestioning reverence for authority and chain of command, mimicking military discipline, and many of his fellow students played at being soldiers in the yard during breaks, marching in parades to the beat of war drums. Einstein wanted absolutely nothing to do with this. He once burst into tears watching the other children at their military games, so oppressed did he feel by their ugliness. His disenchantment with religion had only made him less amenable to arbitrary authority, learning by rote, and mechanical discipline. "Suspicion against every kind of authority grew out of this experience," he wrote later in life, "an attitude which has never again left me."

Einstein did not openly push back against the authority of his teachers—he was in school to learn, not to lead a revolution—but his teachers sensed both his superiority of mind and his quiet contempt for their martinet authority. One of his instructors insisted that he was interfering with class discipline merely because he was sitting quietly at the back of the classroom, smiling to himself.

Adding to the tension he was suffering at school, a terrible blow came in the form of a sudden financial and business disaster for Einstein's family. In 1894, Jakob and Hermann Einstein's gas and electric company went under after having been successful and competitive for a number of years. The brothers had mortgaged their homes to support the business, so when they lost a number of important contracts to competing companies, their losses were catastrophic. Einstein was 15 at the time. His family home was sold and torn down, and parents, sister, and uncle went to Italy to try and restart their company. Einstein, however, had to remain behind, boarding with a distant relative, so that he could finish his schooling. Not long after his family's departure, however, Einstein was diagnosed with nervous exhaustion by Max Talmud's older brother, who was also their family doctor. There is some debate as to whether Einstein pushed for this diagnosis so that he would have an excuse to leave the school where he was so miserably unhappy, or whether the same teacher who had declared that his smiling disrupted the discipline of his classroom had placed pressure on him to withdraw. Either way, Einstein was undoubtedly relieved to see the last of the school. Unbeknownst to his family, he boarded a train for Italy and presented himself at his parents' front door, declaring that he never intended to go to back to Germany, and that he would study at home to prepare for the university entrance exams. Furthermore, he asked his father to help him file paperwork to officially renounce his German citizenship. In just two years, he would reach the age at which German boys were conscripted into the army for mandatory military service, a prospect that filled him with terror. Only by shedding his German citizenship could he escape.

Einstein assisted his uncle in the technical work of his family's gas and electricity company, and studied physics in the mean time. He was two years younger than the required age for admitting students to the Federal Institute of Technology in Zurich, but he had no difficulty acquiring letters of recommendation from those who knew of his abilities, persuading the college authorities to let him take the examinations in spite of being only fifteen. Despite all his preparations, Einstein failed the entrance exam. The mathematics and physics examinations naturally posed no difficulty for him whatsoever, but because he was a foreigner and, technically, a high school drop out, he was required to answer questions on a wide range of subjects having nothing to do with scientific or technical studies, such as literature, French, and politics. But his performance in the math and science sections were so impressive that the examiners encouraged him to enroll in a Swiss high school; graduation from such a school would automatically guarantee admittance in to the polytechnic school the following term.

Einstein accordingly moved to the village of Aarau in Switzerland, and where he enrolled for a final year of school, boarding with the large, kind family of one of the school's teachers, Jost Winteler. Much to his relief, the Swiss school was a much better fit for his personality and talents than his militaristic Munich school had been. As Walter Isaacson describes it,

"It was a perfect school for Einstein. The teaching was based on the philosophy of a Swiss educational reformer of the early nineteenth century, Johann Heinrich Pestalozzi, who believed in encouraging students to visualize images. He also thought it important to nurture the "inner dignity" and individuality of each child. Students should be allowed to reach their own conclusions, Pestalozzi preached, by using a series of steps that began with hands-on observations and then proceeded to intuitions, conceptual thinking, and visual imagery. It was even possible to learn—and truly understand—the laws of math and physics that way. Rote drills, memorization, and force-fed facts were avoided."

Einstein was very happy attending this school and living with the Winteler family. Jost Winteler became his mentor, and he addressed Mrs. Winteler as "Mama." Furthermore, he began a romance with the oldest Winteler daughter, Marie, who was eighteen to his sixteen. His grades, at the end of the school year, were the highest possible in every subject except for French (he had been forced to take remedial tutelage in chemistry and French throughout the year; he soon caught up in chemistry, but never quite managed the French.) Overall, he was second in his class. Soon, he was scheduled to take the rather less overwhelming version of the technical college entrance exams given to graduates of the local high school. For his French essay—again, his weakest area of performance— he wrote an interesting statement on his hopes for his future career:

"If I am lucky and pass my exams, I will enroll in the Zurich Polytechnic. I will stay there four years to study mathematics and physics. I suppose I will become a teacher in these fields of science, opting for the theoretical part of these sciences. Here are the reasons that have led me to this plan. They are, most of all, my personal talent for abstract and mathematical thinking... My desires have also led me to the same decision. That is quite natural; everybody desires to do that for which he has a talent. Besides, I am attracted by the independence offered by the profession of science."

The Zurich Polytechnic, where Einstein enrolled in 1896, was a combination of a modern engineering institute and a teacher's college; it could not grant doctoral degrees, but it prepared its students to teach math and science in secondary schools. Despite the fact that he had always shown a marked facility for mathematics, math interested him less than physics, and his marks in mathematical subjects were always slightly lower than his top-ranked physics grades, particularly in geometry. Later in life, he admitted that he had not fully appreciate the interdependency of mathematics and physics, and when he was working on his gravitational theories he was forced to seek the help of one of his old geometry professors in order to work out the mathematics in his paper. Einstein was always more attracted to theory than to practicalities. Around the time of his admittance to the polytechnic school, his father and uncle's second business in Italy failed. They dissolved their business, and Jakob Einstein went to work as an engineer for a larger firm, while Hermann Einstein decided to start yet another power supply company. His hope was that his son would come to work for him when he finished his education, but Einstein was dubious about his father's prospects for further success in the same industry where he had suffered repeated failures; furthermore, he could not face the idea of a career in engineering. He wrote to a friend that he wished to think for the sake of thinking, as music is played for the sake of music, rather than apply his scientific knowledge to commercial goals for the sake of making money.

Einstein was not entirely excused from intellectual disappointments during his time at the polytechnic school. His favorite physics teacher, Heinrich Weber, was very sound on the history and origins of physics, but in Einstein's opinion he did not pay sufficient attention to recent developments and breakthroughs in the field. Einstein's disdain irritated Weber, who remarked, "You're a very clever boy, Einstein. But you have one great fault. You'll never let yourself be told anything." Another of his physics professors actually suggested that he apply his talents to law or medicine; in his opinion, Einstein did not have the necessary head for a career in physics. Einstein was particularly irritated that Weber did not teach any of the recent breakthroughs in theoretical physics pioneered by James Maxwell. The physicist James Maxwell is perhaps most famous in popular culture for his thought experiment about a potential violation of the second law of thermodynamics, known informally as "Maxwell's Demon". A professor at Auburn University explains Maxwell's Demon thus:

"Suppose that you have a box filled with a gas at some temperature. This means that the average speed of the molecules is a certain amount depending on the temperature. Some of the molecules will be going faster than average and some will be going slower than average. Suppose that a partition is placed across the middle of the box separating the two sides into left and right. Both sides of the box are now filled with the gas at the same temperature. Maxwell imagined a molecule sized trap door in the partition with his minuscule creature poised at the door who is observing the molecules. When a faster than average molecule approaches the door he makes certain that it ends up on the left side (by opening the tiny door if it's coming from the right) and when a slower than average molecule approaches the door he makes sure that it ends up on the right side. So after these operations he ends up with a box in which all the faster than average gas molecules are in the left side and all the slower than average ones are in the right side. So the box is hot on the left and cold on the right. Then one can use this separation of temperature to run a heat engine by allowing the heat to flow from the hot side to the cold side."

Maxwell devised this thought experiment in a letter in letters to physicists Peter Guthrie Tait and John William Strutt in the late 1860's and 1870's. Whether Einstein was familiar with this particular aspect of Maxwell's work is unknown, but he certainly longed for a more thorough explanation of Maxwell's theories, and theories by other theoretical physicists who had done more recent work in physics, than he was receiving in the polytechnic school's curriculum.

Einstein's entrenched disenchantment with authority and received wisdom made him, in some ways, an even more problematic college student than he had been at his militaristic Munich high school. Having little interest in mathematics, he frequently skipped his lectures with one of the world's leading mathematicians, Hermann Minkowski, which was somehow ironic as Minkowski's theories about space-time would later supply the mathematical underpinnings of Einstein's own theories of relativity. Minkowski later wrote that Einstein's revolutionary work in physics came as a shock to him, because he had been such an indifferent and lazy student in school. Einstein went from making the top grades in his year at the mid-point of his education to failing at least one physics practicum and coming last out of four students taking the final exams for graduation in 1900. When presented with instructions for performing lab experiments, he was known to toss the instructions in the trash and perform the experiment according to his own methods. He almost always came up with the right answers, but his contempt for established procedure irritated his instructors profoundly, and on at least one occasion he made a dangerous misstep that resulted in an explosion, damaging his right hand and leaving him unable to write or play the violin until he had fully recovered. But then, physics is divided between the theoretical and experimental fields, and Einstein's heart was always in theory.

Socially, Einstein was still something of a loner; because he felt that his physics instructors were neglecting the most recent advancements in the field, he devoted a great deal of his time to acquiring the works of leading contemporary physicists and studying them in private, just as he had done with the borrowed math and science textbooks he received from Max Talmud as a young boy. But he did have the capacity to make friends, and some of the social connections he formed in Zurich remained with him through his entire life. One of the friends he made at the polytechnic, Marcel Grossman, was the son of a Jewish factory owner and supremely talented in mathematics; the notes that he took in the mathematics courses he shared with Einstein were almost solely responsible for getting Einstein through his examinations. With Grossman, Einstein began visiting coffee shops and evening parties in Zurich, and when Einstein was struggling to find work after graduation, Grossman's influence would be responsible for getting Einstein his job at the patent office, where he wrote his first papers on relativity.

Music also helped ease Einstein into social gatherings. Some of the evening coffee and dinner parties he attended were also musical soirees, where his violin performances were much sought after. His favorite composers, indeed, almost the only composers he played, were Mozart and Bach. He saw in their compositions a reflection of his feelings about the physical universe. He once remarked that when he listened to Beethoven, he was conscious of an artist creating music, but that when he listened to Mozart, he felt as if the composer had plucked his melodies from the fabric of the universe, where they already existed.

Einstein finished his studies at the Zurich Polytechnic in 1900, with, as has already been mentioned, indifferent success in his examinations. Clayton Gearheart writes the following description of the examinations Einstein had to pass, and differentiates them from the sort of exams offered by traditional European universities, as well as contemporary American universities:

"Now the ETH, as we have seen, was not a university but a technische Hochschule, more oriented toward practical degrees and professional certification; and the Swiss system may have differed slightly from the German. In any case, Einstein was actually given grades in a full one-quarter of the courses listed on his final transcript! (I am not sure how these grades were determined; certainly neither Einstein nor any of his biographers mentions examinations in the courses.)

Einstein did take an oral intermediate examination in 1898, and the oral graduation examination in 1900 that I have already described. The latter was required for certification as a secondary school teacher—that is, it was more in the nature of a professional certification exam. Such was the horrible gamut of examinations of which Einstein complained.

"One can only speculate how Einstein might have done in an American university, with its frequent examinations. One American academic, having heard Einstein's story, responded with horror that at his own university,

"'Einstein would never have made the Dean's List . . . I also doubt very much if he could have passed the college entrance examinations, and he would probably have been put on probation.'"

Mileva Marić

When Einstein graduated from his Swiss cantonal high school and entered Zurich Polytechnic, he left behind the loving and generous support of the Winteler family, with whom he had boarded for over year. During his time in the Winteler home, he had developed a romantic relation with Marie Winteler, the oldest daughter, who was two year his senior and a trained schoolteacher. The relationship was a serious one, despite the fact that they were both teenagers. At Einstein's encouragement, Marie Winteler had begun writing letters to his mother, and both families regarded the couple as engaged, or as near to it as a sixteen year old and eighteen year old who were still finishing their educations could be. Einstein's mother was very fond of Winteler, and Winteler's family regarded Einstein as a member of the family.

Einstein wrote faithfully to Winteler when he first entered the polytechnic school. He also sent her bundles of his laundry for Marie to wash and return by mail. Gradually, he began neglecting to send letters with his dirty clothes, and eventually he ceased communicating with her entirely. At first, Winteler put this down to his absentmindedness; she wrote to his mother who was of the opinion that Einstein had merely grown lazy. But as time passed, Einstein was forced to admit that he no longer had the same feelings for Winteler; he wrote to her parents that he could not justify returning to Aarau for a visit with their family, as it could only bring Marie more pain.

The dissolution of Einstein and Winteler's relationship was due to several factors. Firstly, he was still a teenager, a period which, in anyone's life, can involve radical fluctuations of emotion and passion. Secondly, while Winteler had remained behind in a familiar home environment where little in her life changed from day to day, Einstein had been plunged into a new world, with new friendships, new influences, and new opportunities to shape the direction of his adult life; as such, it is not terribly surprising that the charms of his old life began to pale in comparison to the new vistas opening before him.

Thirdly, Einstein had met someone else: a young woman by the name of Mileva Marić, who, in the fullness of time, would become his first wife.

By the time she met Einstein when she was twenty one, Marić had already had an extraordinary career for a young woman at the turn of the twentieth century. She was the oldest child of a Serbian peasant who had married a woman considerably more wealthy than he. Milos Marić was immensely proud of his daughter's evident brilliance, and he was determined that, with his support, she would gain admittance to the all-male schools that would allow her to pursue her studies in mathematics and physics. In 1892, he convinced the Royal Classical High School in Zagreb to let her study as a private student, and when she took her final examinations she gained the school's top-ranked marks in physics and mathematics. She began studies at the Zurich Polytechnic school the same year as Einstein; in a class of six people, she was the only woman— indeed, the only woman in the whole school.

It seems that Marić and Einstein became friends shortly after their meeting, but that their relationship was strictly platonic for the first year or so. Apparently, however, Marić began to perceive that she was developing feelings for

Einstein—feelings that could only distract her from her studies. It isn't surprising that Marić was dubious of the wisdom of beginning a romantic relationship with a fellow student. Even for women who had careers, marriage was generally expected to put an end to them, as children were expected to follow shortly, and to absorb all of their time and energy. Even schoolteachers generally quit their jobs as soon as they found husbands. Marić chose to distance herself from Einstein by moving to Germany; at Heidelberg University, one of Europe's oldest and most famous institutions of higher learning, she was not permitted to enroll as a degree-seeking student, but she was permitted to audit classes.

From Heidelberg, Marić answered the letters Einstein wrote to her—some of them unusually long letters, at least by Einstein's standards—but there is little flirtation or perceptible romantic intent in them, though she reflected fondly on the long hiking tour they took together in the Alps. She devoted far more of her writing energy to recounting the lectures and ideas she was encountering in her studies, and her letters discussed overarching concepts about life and the workings of the universe far more than the personal concerns of human beings—which was rather Einstein's style. As their friendship deepened through their correspondence, Einstein began encouraging her to return to Zurich, and April of 1896 she did so. Upon her return, she took rooms in a boarding house down the street from Einstein's rooms, and it seems that shortly after this they began regarding themselves as something of a couple, often walking together, talking, exchanging books, and reading in each other's company.

It may come as something of a surprise to the modern reader, who is mostly familiar with photographs of Einstein taken after he moved to the United States, when his hair was grey, wild, and unkempt, that in his youth Einstein was considered an extremely handsome young man. He had short dark hair, a neat dark mustache, height, good health, and regular features, and he was generally considered to be a first-rate specimen of the style of male handsomeness that was in vogue in Europe in the late nineteenth century. He was a bit careless of grooming, and his personal habits were considered a little eccentric, but his friends regarded him as the sort of handsome man who could win the affections of any woman he chose to be interested in. By contrast, Mileva Marić was not considered to be a particularly attractive woman. She was small, slender, and delicate, with intense dark eyes and serious, unsmiling mouth; furthermore, she was sometimes in poor health. She limped due to a hip displacement she had suffered as a child and she was sometimes subject to attacks of tuberculosis. Judging by photographs, these estimations of Marić's supposed unattractiveness seem quite ungenerous—by modern standards, her face seems interestingly pretty. Yet Einstein's mother, who had been so fond and approving of his relationship with Marie Wenteler, had a somewhat dubious reaction upon being presented with her photograph. Einstein's friends found it baffling that he should be interested in such a woman, apparently disregarding the profound attraction he must have felt towards a woman with whom he could discuss physics as an equal.

But as Einstein biographer Walter Isaacson writes,

"It is easy to see why Einstein felt such an affinity for Marić. They were kindred spirits who perceived themselves as aloof scholars and outsiders. Slightly rebellious toward bourgeois expectations, they were both intellectuals who sought as a lover someone who would also be a partner, colleague, and collaborator. "We understand each other's dark souls so well, and also drinking coffee and eating sausages, etcetera," Einstein wrote her."

"How proud I will be to have a little Ph.D. for a sweetheart," he also wrote. It is evident that Marić intellectual talents exerted an unshakeable influence over Einstein's romantic imagination.

In what may be interpreted as further evidence of Einstein and Marić's similarity and compatibility, when the two of them sat for their examinations, they received the two lowest marks in their class. Marić, in fact, did not even qualify for a teaching diploma. She decided to take the exams again the following year, and she was making plans to convert her graduating thesis into research for a Ph.D.

After graduating, Einstein left Zurich and met up with his family, of whom he had seen little since beginning his studies, and retreated with them to a house in the mountains, for a long summer vacation that would avail him of the opportunity to recover from the stress of his examinations.

Chapter Two: Young Einstein

Domestic Conflict

Having returned to the company of his family and parents, Einstein soon found his mother making tentative inquiries into "the Dollie affair", as she put it (Dollie was Einstein's nickname for Mileva Marić and he referred to her thus in his letters home.) Einstein—seemingly from a wish to communicate to his parents that he had matrimonial intentions towards Marić, or else possibly in a gentle attempt to communicate that he and Marić were sexually active—referred to her as "my wife". His mother promptly flew into a passion, throwing herself down on the bed and weeping profusely, until a visit from an elderly neighbor forced her to regain his composure.

With rather more honesty than tact, Einstein related to Marić in his letters how entrenched was his family's opposition to their relationship. But what he lacked in delicacy he made up in vehemence. His natural resistance to authority and what he perceived as the petty morality of the bourgeois class (at least, it seems, as it pertained to the ban on pre-marital sex) caused him to dismiss his family's objections, and he declared to Marić that until he met with such protests, he did not realize how in love with her he truly was.

Apparently, Einstein's parents were unbothered by the fact that Marić was not Jewish, nor they did object strenuously to her Serbian ethnicity— rather, they seemed to feel that she was unfit to be their son's wife on the grounds that she was unwomanly, "a book", in his mother's words, presumably referring to her intellectual pursuits, when what Einstein needed was "a wife". Being three years Einstein's senior apparently made her far too old—his mother claimed that by the time Einstein was thirty, his wife would be "an old witch". Of course, Marie Westeler, of whom Einstein's mother was thoroughly approving, was two years older than he, but her other womanly qualities apparently made up for this. They also objected to the fact that she was not in robust health—added to their fears of an unplanned pregnancy was the concern that she would break Einstein's heart by dying prematurely. As Einstein puts it,

"Papa has written me a moralistic letter for the time being, and promised that the main part would be delivered in person soon. I'm looking forward to it dutifully. I understand my parents quite well. They think of a wife as a man's luxury, which he can afford only when he is making a comfortable living. I have a low opinion of this view of the relationship between man and wife, because it makes the wife and the prostitute distinguishable only insofar as the former is able to secure a lifelong contract from the man because of her more favorable social rank. Such a view follows naturally from the fact that in the case of my parents, as with most people, the senses exercise a direct control over the emotions. With us, thanks to the fortunate circumstances in which we live, the enjoyment of life is vastly broadened. But we mustn't forget how many existences like my parents' make our existence possible. In the social development of mankind, the former are a far more important constituency. Hunger and love are and remain such important mainsprings of life that almost everything can be explained by them, even if one regards the other dominant themes. Thus I am trying to protect my parents without compromising anything that is important to me — and that means you, sweetheart!"

It must have struck Einstein as ironic that his parents suddenly felt that his marrying would constitute an unjustifiable luxury, considering how enthusiastically they had encouraged the prospect of his union with Marie Westeler.

Einstein's letters reveal a certain assumption that because Marić was an intellectual, and their relationship founded on the rarefied pursuit of theoretical physics rather than the mundane romantic gestures that had defined his relationship with Marie Westeler, she was above such petty, ordinary female concerns beauty, tenderness, and to a certain extent, feelings. This may account for the nearly brutal frankness with which he recounted to her how much his family disliked the mere thought of her. It was as if Einstein subscribed to the old prejudice that brains and beauty cannot coexist in women, and that because Marić knew herself to possess brains, she could not possibly care in the least if no one supposed her beautiful. What Marić thought of Einstein's insensitive teasing on this score is difficult to determine. His letters from this period also relate a certain condescending appraisal of his mother's intellect, as well as his father's—he supposes that they could not possibly enjoy life as much as people like himself and Marić, because their thoughts and concerns were limited to the petty domestic, social, and financial worries that preoccupy ordinary minds. But, he allows, not everyone in society can be a genius, and limited persons such as his parents were necessary to permit the world to function properly.

The limitations of Einstein's father in particular had led to the failure of two different business ventures since Einstein was a child, and his third business had come perilously close to collapsing just as Einstein was finishing school. He often expressed the wish that his father would simply take up a salaried position in a large firm, as his uncle Jakob had done. While residing in his family's vacation lodgings, playing violin at his mother's request to amuse and charm to other guests, Einstein was drawn into discussions with his father about the employment he would pursue now that his education had been completed. His father was not with the family, as he was still busy at work in Milan. Einstein decided to visit him personally, both in an attempt to quell his father's unease over his relationship with Marić, and, as he put it, to learn something about his father's business so that he could take over management of the company in an emergency.

His father, unsurprisingly, was very glad to see him. Einstein took his duties as his parents' only son very seriously, and he applied his attention diligently when his father took him on tours of his company. They had at least one argument about Marić, but when Einstein threatened to cut his visit short, his father relented and dropped the subject. While Einstein had no interest in becoming an engineer, he certainly had the training for the job, and he was reluctantly prepared to take the job up if necessary, or if his father pressed him to do so—after all, he was unemployed, and his job prospects were not as bright as they might have been if he had scored higher on his examinations. Fortunately for the field of theoretical physics, however, Hermann Einstein remained in good health, and his son was not forced to make that choice.

Employment

Einstein wished from the beginning to obtain an academic job, but the difficulty he had in finding one is rather astonishing, at least for contemporary readers who know of his later accomplishments. It was typical for students at Zurich Polytechnic to obtain a position as an assistant to one of their former professors at the school, and Einstein accordingly applied for such a position. However, his comparative laziness as a student and his disdain for authority had not made a favorable impression on his former instructors and prospective employers, and none of them were interested in hiring him, even when no other former students were available to take the position.

Back in Zurich, Einstein was living in Mileva Marić's apartment and supporting himself as best he could by taking on tutoring jobs in mathematics. The lack of a steady job was of great concern to him, not only for financial reasons. Ever since he had, with his father's help, renounced his German citizenship as a fifteen year old, he had been stateless, officially a citizen of no country. He was now ready to apply for Swiss citizenship, however, and he was required to have steady employment in order to be considered. When the hoped-for job at the polytechnic school proved a bust, he wrote on his citizenship papers that he was giving mathematics lessons until he could gain more regular work.

In the mean time, Einstein was spending time reading and studying with Marić, and working on his first theoretical papers, which led to his first publication, an essay on capillarity. Capillary action refers to the tendency of molecules to be attracted to one another, as when water climbs through a straw; the principle of capillarity was not yet a completely established and accepted aspect of physics at this time, nor were the existence of molecules and atoms, which shows some willingness on Einstein's part to espouse unconventional theories when they suited his ideas of the universe. This particular paper was not especially inspiring, particularly when compared to his later writings, and the fundamental theory was later proven to be erroneous. However, he was published in the German science journal *Annalen der Physik* (Annals of Physics) in 1900, which was not an insignificant accomplishment for a teenager who had just finished school. It also bolstered his credentials in seeking a job— having been turned down for an assistant's position by his own former professors, he was now soliciting professors in universities all over Europe now.

In February of 1901, Einstein finally gained his Swiss citizenship—a good thing, as he would not be able to be hired for any of the teaching positions he was seeking unless he was a Swiss citizen. The examination process was stringent, but Einstein's enthusiasm for his adopted homeland was genuine—so genuine that, after having taken extraordinary measure to evade conscription into the German army as a 17 year old, he presented himself for inspection at the Swiss conscription office without any apparent hesitation. (He was declared unfit for service on grounds of having sweaty feet, flat feet, and varicose veins.) However, around the same time his citizenship papers were issued, his parents demanded that he return to Milan, unless he had already found a job. They were, Einstein was convinced, making this demand in an effort to put distance between him and Mileva Marić. But as he remained unemployed and financially dependent on his parents, he had no choice was to return to Milan as they demanded.

His father, however, was as supportive as he knew how to be in all aspects of life except those pertaining to Marić. He was deeply concerned that his Einstein's career should progress, and it disheartened him nearly as much as Einstein himself when so many applications for employment to professors in the universities of Europe went unanswered. Unbeknownst to Albert, Hermann Einstein took it upon himself to, rather awkwardly but with the most sincere of fatherly feelings, write to one particular professor to whom Einstein had sent a copy of his article on capillary action.

"Please forgive a father who is so bold as to turn to you, esteemed Herr Professor, in the interest of his son. Albert is 22 years old, he studied at the Zurich Polytechnic for four years, and he passed his exam with flying colors last summer. Since then he has been trying unsuccessfully to get a position as a teaching assistant, which would enable him to continue his education in physics. All those in a position to judge praise his talents; I can assure you that he is extraordinarily studious and diligent and clings with great love to his science. He therefore feels profoundly unhappy about his current lack of a job, and he becomes more and more convinced that he has gone off the tracks with his career. In addition, he is oppressed by the thought that he is a burden on us, people of modest means.

Since it is you whom my son seems to admire and esteem more than any other scholar in physics, it is you to whom I have taken the liberty of turning with the humble request to read his paper and to write to him, if possible, a few words of encouragement, so that he might recover his joy in living and working. If, in addition, you could secure him an assistant's position, my gratitude would know no bounds. I beg you to forgive me for my impudence in writing you, and my son does not know anything about my unusual step."

Einstein was unsuccessful in his job search until his old friend from Zurich Polytechnic, Marcel Grossman, wrote to tell him that an inspector's position had opened up in the Swiss Patent Office. Grossman's father, the wealthy businessman, knew the director of the office, and was offering, at his son's encouragement, to use his influence to secure the position for Einstein. It wasn't precisely the kind of work Einstein had dreamed of, but it was quiet, non-commercial, and scientific in nature, and he had despaired of finding work for so long that he greeted the prospect of such employment with the deepest joy.

Why it was that only by the personal intervention from the father of a friend could Einstein getting a job in a science-related field is a baffling mystery to anyone who knows of his later work. Einstein himself had several theories as to what might be working against him. Firstly, he believed that one of the physics teachers at the polytechnic who had a special dislike for him was giving negative references when prospective employers inquired about Einstein. Secondly, Einstein felt that anti-Semitism was at least partly to blame; on that account, he hoped, having moved to Italy to appease his parents, that his job search would be more successful there. He perceived anti-Semitism as being potentially less pervasive in Italy than it was in Germany.

Marić's Pregnancy

The patent office position would not be available for some months yet, so Einstein continued looking for employment in the mean time. At length, he found a teaching position in a technical school, covering for another teacher who had been on leave. But then when the summer of 1901 arrived, he was eager for a vacation, and even more eager to be reunited with Marić, though they had scarcely been separated a month at this point. Einstein proposed a vacation to Lake Cuomo, in the mountains.

Marić and Einstein enjoyed their vacation very much, but an inconvenient circumstance came about as a result of their time together: Marić discovered that she was pregnant. This was not altogether the disaster that it might have been; birth control, including abortifacient drugs, were available in Zurich, and pregnancy out of wedlock was less uncommon than one might think. It seems likely that Marić, however startled by her pregnancy, came to see it as a good thing, and Einstein, likewise, took a seemingly lighthearted tone in inquiring after the developing child in his letters. He promise Marić that he would get a job immediately, any job that he could, no matter how distasteful or demeaning, and that once he was employed they would marry. He would not hear of getting married before that happened, however.

Marić was living in Zurich because she had been preparing to re-take her examinations at the polytechnic school; her goal was to earn a Ph.D. afterwards. But the stress of her secret pregnancy must have divided her attention from her studies, because she made the same grade on her second exam attempt as she had on her first, a 4.0 out of 6. It did not constitute a passing grade. This failure must no doubt have been crushing to Marić, who had fought so hard just to gain access to an education; not only would her failure be held against women seeking higher education in the sciences to come after her, it would make all those who had been dubious about allowing her to study in the first place feel justified in their reluctance. Worse still, her father had devoted enormous time, effort, and money to her education. For Marić to be forced to return home and confess an unplanned, premarital pregnancy on top of her failure to get a degree and the consequent ruin of her hopes for a scholarly career, must have been devastating.

Einstein applied for more teaching jobs after his substitute job ended and he continued to get tutoring positions for low pay. He was hired by a private school to be a private tutor to an English boy from a wealthy family for one hundred and fifty francs a year. In the mean time, he wrote to Marcel Grossman again and reminded him that he was waiting to hear about the job in the patent office. Luckily for Einstein, Grossman wrote back, advising him that the job was about to be advertised publicly, but that he was certain to be selected as soon as he applied. The timing was fortuitous, as Einstein's lifelong difficulty in submitting to authority was causing him to clash with the head of the private school where he was working.

Marić was naturally eager to be married as soon as possible—she and Einstein could not meet publicly while her pregnancy was visible, so she was forced to wait on a wedding with an uncertain date before she could begin making serious plans for her future. Both prospective parents were concerned about the stigma of illegitimacy if the child was born before they were married. They acknowledged the possibility that the baby would have to be given up for adoption, an outcome that neither of them were happy about. Einstein, who initially had referred to the unborn child as "the boy", had changed his mind and decided that the baby was probably a girl, and thus began to refer to her as "Lieserl".

Unlike Marić's family, Einstein's knew nothing about the pregnancy. It was quite understandable that he should have kept it a secret from his parents, considering how his mother had predicted that pregnancy and disgrace would be the inevitable consequence of his unconventional relationship with Marić. Their opposition to Marić continued unabated; Pauline Einstein even went so far as to write to Marić's parents, declaring that the union was very much against their wishes and that they wanted nothing to do with her. Hermann Einstein signed the letter as well. This was a source of considerable stress to Marić, who wondered how anyone could be so "wicked" as to wish to embitter the life of her own son and a woman they had never met.

Lieserl

Whether the elder Einsteins liked it or not, however, Einstein was looking forward to the marriage which would be made possible as soon as he started receiving a salary from the patent office. He wrote to Marić, expressing his hope that they would not, once they were married, settle into the ordinary, boring married life of other middle class couples—it was very important to him that Marić should continue to be his "witch" and "street urchin" (nick names that he gave her because of her small size and vaguely childlike appearance) and that their lives should continue to revolve around reading physics journals together and discussing scientific concepts late into the evening. It is possible that Einstein did not fully appreciate the disruption that an infant would pose to this cozy domestic routine, but in any case, it was a fondly cherished dream that sustained him through long months of being unemployed and separated from Marić.

As it happened, Einstein and Marić were unable to marry prior to the birth, and Marić was living at home with her parents when their daughter, Lieserl, was born. The birth was long and taxing and Marić faced a long recovery afterwards. Einstein had to depend on letters from her father for details about the child: how big she was what she looked like, whether she was in good health.

After these first few letters, however, Lieserl Einstein drops almost entirely out of the historical record. She was not, in the end, raised by Einstein and Marić; in fact, according to Einstein biographer Walter Isaacson, there is no indication that Einstein himself ever met his first child. Secrecy surrounded her birth—if word had got out that Einstein had an illegitimate child, even with a woman whom he married immediately after, he would lose the job in the patent office on which all of his hopes for the future depended. Apart from the first flurry of letters written after Lieserl's birth, which were concealed by Einstein's family until thirty years after Einstein's death, none of Einstein or Marić's letters make any reference to her. Her existence was entirely unknown to the public until the 1980's. It is possible that she had congenital defects or learning disabilities, and that she either died very young or that she was given to one of Marić's close friends to raise, who later adopted her and changed her name. What became of her ultimately, no one is certain; apart from the announcement of her birth, and the fact that she contracted scarlet fever at the age of nineteen months in 1903, no further details are definitely known about her.

Chapter Three: Einstein's Early Career

The Patent Office

On June 16, 1903, Einstein was hired "provisionally as a Technical Expert Class 3 of the Federal Office for Intellectual Property with an annual salary of 3,500 francs," a tolerably decent salary for a young man in his first professional job—better money than he would have made in any of the assistant academic positions he had applied for. Einstein's duties were fairly straightforward: people submitted designs for their inventions to the office, hoping to receive an official Swiss government patent that would protect the design as their exclusive property. Einstein examined the designs in order to determine whether the machines would function as the inventor claimed they would function. It was also his job to make certain that the inventor had not simply copied someone else's designs.

Einstein had been warned by other employees of the patent office that his job was boring and carried little prestigious—he would be the lowest ranked of all the other employees. But Einstein scoffed at this, saying that, to banal minds, anything could be boring. In fact, Einstein's engineering and other scientific training made the work of evaluating patents very easy, but he did not find it boring in the slightest. It was, he told Marić, work "uncommonly diversified". More to the point, Einstein was able to complete his day's work so quickly that he had many hours left over during which his presence at the office was required, but he was free to work on his own theories and papers on topics related to physics. (It appears that his supervisor was not unaware of this extracurricular use of Einstein's time on the job, but as Einstein never failed to complete his work in a thorough and timely manner, he did not complain.) Altogether, the excellent pay, interesting work, and free time made the patent office job ideal employment for Einstein. He came to feel that he was much better off in this position than he would have been had he been successful in obtaining an academic position. As Walter Isaacson writes,

"Had he been consigned instead to the job of an assistant to a professor, he might have felt compelled to churn out safe publications and be overly cautious in challenging accepted notions. As he later noted, originality and creativity were not prime assets for climbing academic ladders, especially in the German-speaking world, and he would have felt pressure to conform to the prejudices or prevailing wisdom of his patrons. 'An academic career in which a person is forced to produce scientific writings in great amounts creates a danger of intellectual superficiality,' he said."

Einstein was to work happily as a patent inspector for seven years, until after he had published the papers on relativity that revolutionized physics, at which point he was finally offered a professorship at a university.

While he worked as a patent examiner, Einstein lived and worked in the Swiss city of Bern, in an office in the Postal and Telegraph Building. Every day, on his way to work, he passed by an enormous clock tower, an historic landmark in Bern, and all the government and train station clocks in the country were synchronized with its time-keeping. This clock tower is regarded today as something of a monument to the work Einstein did while he lived in the city, the inspiration that provoked him to think about time in such an original way.

The Olympia Academy

Before he obtained his job in the patent office, Einstein had placed the following advertisement in the local newspapers in the hopes of attracting paying pupils:

"Private lessons in mathematics and physics for students and pupils given thoroughly [by] Albert Einstein, owner of the Swiss polyt. subject teacher diploma, Gerechtigkeitsgasse 32, 1. floor. Trial lessons for free."

Seeing this ad, a young Romanian man four years older than Einstein named Maurice Solovine decided to pay him a visit. He was not a student in the traditional sense of the word, and he certainly wasn't the sort of pupil that Einstein had expected to attract in advertising his tutoring services, but Einstein found Solovine interesting to talk to. Solovine had not made up his mind what sort of career he wished to pursue—he was something of an autodidact and had studied philosophy, physics, and other subjects on his own—and Einstein told him, "It is not necessary to give you lessons in physics, the discussion about the problems which we face in physics today is much more interesting; simply come to me when you wish, I am pleased to be able to talk to you." Solovine suggested that they should read serious books, and meet to discuss their reading. Shortly after they began, they drew into their circle another graduate of the Zurich Polytechnic school, Conrad Habicht, who was now pursuing a career as a mathematician.

These three semi-serious scholars named themselves "The Olympian Academy", a lofty sounding title that was intended as something of a joke, mimicking similar academic circles in universities that took themselves much more seriously. Einstein was named president of the "Academy" on the basis of the fact that it was his apartment that hosted most of the meetings. They generally began meetings by eating dinner—simple meals of sausage, cheese, and fruit. Einstein would later be notorious for his indifferent palate; he never seemed to notice what he was eating, so long as it satisfied his hunger. This deficiency of taste was made evident when Habicht and Solovine pooled their limited resources to add caviar to the dinner offerings as a surprise for Einstein's birthday. They waited expectantly for him to comment on the expensive luxury, but he didn't seem to notice. When they drew his attention to it, he was surprised and a little embarrassed. "For goodness' sake," he said, "So that was the famous caviar! Well, if you offer gourmet food to peasants like me, you know they won't appreciate it."

Together, the members of the Academy read books such as Ernst Mach's *Analyse der Empfindungen (Analysis of the Sensations)* and *Die Mechanik in ihrer Entwicklung (Mechanics and its development),* Karl Pearson's

Grammatik der Wissenschaft (Grammar of Science), Henri Poincarés *Science and Hypothesis*, John Stuart Mill's *Logic*, David Hume's *Treatise of Human Nature*, Spinoza's *Ethics*, and novels such as *Don Quixote* and plays such as Sophocles' *Antigone*. Occasionally, Einstein would play the violin for Habicht and Solovine, or they would talk through the night and go out on a mountain hike after sunrise. Though the Academy would last only a short time, as its members left the city for other jobs and opportunities, the course of reading they pursued together had a profound effect on Einstein's thinking, and his friendship with Solovine and Habicht endured for the rest of his life.

Marriage and Domestic Life

Hermann Einstein, much to the distress of his only son, became seriously ill in 1902, not long after Einstein started work at the patent office in Bern. The elder Einsteins were still resident in Milan, so Einstein went to Italy to be with his family as his father approached his deathbed. Einstein and his father had not always had an uncomplicated relationship; the superiority of Einstein's intellect, his dismissal of authority and the petty domestic concerns that dominated his parents' thinking, his lack of interest in an engineering career in his father's business, and especially his parents' opposition to his relationship with Mileva Marić, all stood in the way of their forming an especially close relationship. Nonetheless, Einstein loved and respected his parents, and had always tried to be a dutiful son, according to his own lights. His father's illness distressed him greatly, and his death came as an enormous blow. However, one good thing resulted from it: as he was dying, Einstein's father finally gave his blessing for his son to marry Marić.

Albert Einstein and Mileva Marić were married in Bern in a civil service attended only by Maurice Solovine and Conrad Habicht; Einstein's mother and sister, as well as Marić's parents and six siblings, all declined to attend. Marić moved into Einstein's flat and began attending meetings of the Olympia Academy, as well as taking up housekeeping duties. But after a honeymoon period of domestic coziness, it seems that tensions began to grow between the new married couple. This is scarcely surprising; their relationship had already undergone a great deal of strain, between the long periods of enforced separation, the hostility of Einstein's family, the strain of concealing a pregnancy, then a child, the physical toll that the birth itself had on Marić's never robust health, and finally the death of Einstein's father. Marić may have been missing her infant daughter very much. It is not known for certain who was raising her, but letters between Marić and Einstein indicate that she had been adopted by someone, because in 1903, when baby Lieserl was nineteen months

97

old, her adoptive family sent word that she had come down with scarlet fever. Marić immediately left Bern to see her.

Scarlet fever is a severe illness caused by the same streptococcal bacteria that lead to conditions such as strep throat, and in the early twentieth century, before antibiotics, it was one of the most common and deadliest of childhood illnesses. Those who survived the disease as children are prone to a variety of serious health conditions as adults, such as kidney disease and rheumatic heart disease. Einstein's few surviving letters (nearly every letter that made any reference to Lieserl's existence were eventually destroyed either by Einstein, Marić, their families, or the family of the person who adopted her) mention his deep concern that her health would be affected by these conditions. Scholars who have investigated the mystery of Lieserl's whereabouts posit that she may have died from the scarlet fever; others theorize that the daughter of Helene Savić, who was Marić's closest friend, was in fact the adopted Lieserl. (Savić's daughter, Zorka, was blind, and blindness is sometimes caused by scarlet fever. Zorka's nephew, who wrote a book about Savić and Marić's friendship, asserts that this was not the case, but declined to produce any evidence to definitively rule out the possibility.)

Whether Lieserl did indeed die of scarlet fever in 1903, or whether it was the pain of being forced to let others raise her, or whether it was something else altogether, Marić was unhappy during most of her marriage to Einstein. She wasn't comfortable in the domestic role she was forced to take up as the wife of a respectable civil servant; though Einstein had declared years earlier that he wanted only for them to continue discussing physics and studying together, someone was required to cook, wash clothes, and clean house, and even if Einstein was prepared to be non-traditional regarding the division of labor, he did work eight hours a day at the patent office. Marić seemed to feel excluded from the intellectual pursuits of the Olympia Academy, though Habicht and Solovine describe her as intelligent and preferring to observe rather than participate.

Considering that Marić had spent her entire life prior to her first pregnancy flouting conventional societal rules for women's pursuits and careers, it can scarcely be wondered at if she found the business of being a wife rather confining and unsatisfactory. And it perhaps possible that Einstein was contributing to her unhappiness; he later asserted that by the time he and Marić were married, he had come to feel some uneasiness at the prospect, but had pressed on regardless from a sense of duty. He had, after all, experienced a change of heart about his relationship with Marie Westeler after leaving Aarau to live in Zurich for awhile; there had been even longer periods of separation in his relationship with

Marić. If his interest was truly waning so early in their marriage, and if, as seems probable considering how tactless he could be, he did not trouble to conceal his disinterest from her, it is hardly to be wondered at that Marić was unhappy.

While visiting Lieserl in Serbia, Marić wrote to Einstein that she had become ill; it turned out, however, that this illness was because she had become pregnant again. She was concerned that Einstein would react poorly, but as it happened, he was delighted—this, after all, was a child he could boast of, whom he could raise as his own and learn to love. On May 14, 1904, their son Hans Albert Einstein was born, and both parents were delighted.

Chapter Four: 1905 The Miracle Year

"There was no sign that [Einstein] was about to unleash an annus mirabilis the like of which science had not seen since 1666, when Isaac Newton, holed up at his mother's home in rural Woolsthorpe to escape the plague that was devastating Cambridge, developed calculus, an analysis of the light spectrum, and the laws of gravity. But physics was poised to be upended again, and Einstein was poised to be the one to do it."

Walter Isaacson, *Einstein, His Life and Universe*

In May of 1905, Albert Einstein wrote the following letter to his Olympian Academy cohort, Conrad Habicht:

"Dear Habicht,

"Such a solemn air of silence has descended between us that I almost feel as if I am committing a sacrilege when I break it now with some inconsequential babble... So, what are you up to, you frozen whale, you smoked, dried, canned piece of soul...? Why have you still not sent me your dissertation? Don't you know that I am one of the 1½ fellows who would read it with interest and pleasure, you wretched man? I promise you four papers in return. The first deals with radiation and the energy properties of light and is very revolutionary, as you will see if you send me your work first. The second paper is a determination of the true sizes of atoms. The third proves that bodies on the order of magnitude 1/1000 mm, suspended in liquids, must already perform an observable random motion that is produced by thermal motion. Such movement of suspended bodies has actually been observed by physiologists who call it Brownian molecular motion. The fourth paper is only a rough draft at this point, and is an electrodynamics of moving bodies which employs a modification of the theory of space and time."

Einstein certainly knew that he was brilliant, and moreover he entirely expected to make essential, even groundbreaking contributions to the field of theoretical physics. However, it is unlikely that he can have appreciated the impact that this letter would have on history, containing as it does a convenient means of proving that the four papers of his that would utterly revolutionize the study of physics were all written between March and June of 1905. For this reason, 1905 is often referred to as Einstein's *annus mirabilis*, or Miracle Year. To use a literary analogy, it is akin to Shakespeare's having written Othello, King Lear, and Macbeth one after another all in 1605, except more so.

Grasping the contents of these four papers can be tricky for those of us who are not scientists, but Richard Panek summarizes it thus:

"Over four months, March through June 1905, Albert Einstein produced four papers that revolutionized science. One explained how to measure the size of molecules in a liquid, a second posited how to determine their movement, and a third described how light comes in packets called photons—the foundation of quantum physics and the idea that eventually won him the Nobel Prize. A fourth paper introduced special relativity, leading physicists to reconsider notions of space and time that had sufficed since the dawn of civilization. Then, a few months later, almost as an afterthought, Einstein pointed out in a fifth paper that matter and energy can be interchangeable at the atomic level specifically, that E=mc2, the scientific basis of nuclear energy and the most famous mathematical equation in history."

Einstein had grown, if not less defiant against authority over the years, at least more pragmatic; he recognized that while the four papers he had just written were groundbreaking, if he ever wanted to get his Ph.D. and switch to an academic career, he could not challenge the conservative notions of the examining faculty too much. So he turned his attention for a time to writing a dissertation on a topic that was traditional and not especially challenging, but that would still illustrate his learning and skills. He decided that the second of his four papers would be the most suitable, and thus his dissertation was entitled "A New Determination of Molecular Dimensions". In the early twentieth century, the process of getting a doctorate was rather different than it is now. Einstein's teaching diploma from the polytechnic degree counted as a university degree (teachers' training then was far, far more rigorous than it is now), but the school did not have the authority to grant doctoral degrees. However, he did not have to enroll in a graduate course at a traditional university to obtain a master's degree, then advance to coursework for a doctoral degree. Because Zurich Polytechnic had close academic ties to the University of Zuric, graduates from the polytechnic school could obtain their doctoral degrees from that institution. All that was required was for Einstein to write a dissertation of sufficient quality under the direction of an advisor and submit it for evaluation to the university faculty.

He was successful in seeking his doctorate, but he continued to work at the patent office for the time being, writing and submitting these earth-shattering papers to the Germany physics journal *Annalen der Physik*. A lifetime of difference experiences shaped and inspired the so called "miraculous" work Einstein did that year, from assisting his engineer uncle in his family's plant as a boy, to a particular daydream he had as a teenager about how travel at light speed would work, as Einstein describes it here:

"If I pursue a beam of light with the velocity c (velocity of light in a vacuum), I should observe such a beam of light as an electromagnetic field at rest though spatially oscillating. There seems to be no such thing, however, neither on the basis of experience nor according to Maxwell's equations. From the very beginning it appeared to me intuitively clear that, judged from the standpoint of such an observer, everything would have to happen according to the same laws as for an observer who, relative to the earth, was at rest. For how should the first observer know or be able to determine that he is in a state of fast uniform motion? One sees in this paradox the germ of the special relativity theory is already contained."

The names of the four papers authored by Einstein during his annus mirabilis year and published in the *Annalen der Physik* are as follows: the first one, on the photoelectric light effect, was called "On a Heuristic Viewpoint Concerning the Production and Transformation of Light"; the second, which described something similar to, but that he did not call, Brownian motion (again, the paper that became his doctoral dissertation) was called "On the Motion of Small Particles Suspended in a Stationary Liquid, as Required by the Molecular Kinetic Theory of Heat". The third paper was the one that changed the entire field of physics forever, the special relativity paper, "On the Electrodynamics of Moving Bodies"; and the fourth, which covered mass-energy equivalence and proposed Einstein's world-famous $E=mc^2$ equation, was called "Does the Inertia of a Body Depend Upon its Energy Content?"

Maurice Solovine, Einstein's friend and fellow Olympian Academy member, described the significance of Einstein's paper on special relativity thus:

"The theory of relativity can be outlined in a few words. In contrast to the fact, known since ancient times, that movement is perceivable only as relative movement, physics was based on the notion of absolute movement. The study of light waves had assumed that one state of movement, that of the light carrying ether, is distinct from all others. All movements of bodies were supposed to be relative to the light-carrying ether, which was the incarnation of absolute rest. But after efforts to discover the privileged state of movement of this hypothetical ether through experiments had failed, it seemed that the problem should be restated. That is what the theory of relativity did. It assumed that there are no privileged physical states of movement and asked what consequences could be drawn from this."

Interestingly, "relativity" was only one of a couple of different names Einstein considered to describe his theory. "Invariance" was another, and he spoke of his "invariance theory" often. It was fellow physicist Max Planck, who discovered Planck's constant and together with Einstein helped intuit all the framework of theoretical physics we rely on today, who settled its destiny by referring to the theory as "relativity" in his own publications.

Hermann Minkowski, one of Einstein's professors at the polytechnic in Zurich, who so famously expressed himself shocked by Einstein's explosion onto the physics scene because he had been so lazy as a student, "decided to give a formal mathematical structure to [Einstein's relativity] theory", according to

Walter Isaacson. Describing Minkowski's efforts, he writes:

"His approach was the same one suggested by the time traveler on the first page of H. G. Wells's great novel The Time Machine, published in 1895: 'There are really four dimensions, three which we call the three planes of Space, and a fourth, Time.' Minkowski turned all events into mathematical coordinates in four dimensions, with time as the fourth dimension. This permitted transformations to occur, but the mathematical relationships between the events remained invariant. Minkowski dramatically announced his new mathematical approach in a lecture in 1908. 'The views of space and time which I wish to lay before you have sprung from the soil of experimental physics, and therein lies their strength,' he said. 'They are radical. Henceforth space by itself, and time by itself, are doomed to fade away into mere shadows, and only a kind of union of the two will preserve an independent reality.'

Einstein, ever neglectful of mathematics, joked that as soon as Minkowski's formulae became interwoven with relativity, he ceased being able to understand his own theory.

Mileva Marić's Contributions

The question of whether Einstein worked with his wife on the four "miracle year" papers, and if so, how far their collaboration extended, has been of keen interest to Einstein's biographers and a curious public in recent decades. Like the fate of his daughter Lieserl, it is one of the bigger mysteries surrounding Einstein's life, and, as with Lieserl, there are only fragmentary indications in a few surviving letters to suggest that Marić may have been a partner in Einstein's work.

It is known more or less for certain that Marić worked on Einstein's papers as an editor. Unsurprisingly, after having produced four papers for the *Annelen der Physik*, plus a doctoral dissertation, on top of working full time, Einstein took to his bed with nervous exhaustion; Maric was left to get his articles in order. And in 1902, he wrote a letter to Marić that said, "How happy and proud I will be when the two of us together will have brought our work on the relative motion to a conclusion!" Marić likewise told her friends that she and her husband were working on an important theory that would make Einstein famous. Einstein himself declared that he relied on his wife to do the mathematical work necessary to his theories, which is entirely plausible, considering how little Einstein liked mathematics. There is also a controversial rumor that Marić was credited by

Einstein in the original version of his manuscripts, and that her name was removed when the papers were being published and disseminated throughout the scientific community.

The question of Marić's contribution level to Einstein's work is a hotly contested one for a number of reasons. It is beyond question that women have historically been excluded from the sciences, and that even when women have won admittance to the world of academic sciences, as Marić did at great personal cost, the achievements of female scientists have been erased from the historical record, often because their male partners take credit for their work, or their contributions have been relegated to the unimportant duties of a lab assistant. For these reasons alone, it is natural to be suspicious that the wife of a scientist as famous as Einstein would have received less credit than she deserved, even if her contributions had been highly substantial. As it is, there is no surviving evidence that indicates that any of the revolutionary ideas Einstein proposed in the four papers from 1905 were actually Marić's ideas, nor did she ever claim that they were. However, it is incontrovertible that she played a critical role in helping Einstein to work his theories out in a way that enabled them to be published. This contribution would have been enough to have cemented her place in the history of physics, had she been a man.

Unfortunately, it seems that Marić was probably right to be fearful for the fate of her academic career when she fled to Heidelberg shortly after first developing a relationship with Einstein. The pregnancy that resulted undoubtedly contributed to her failing her examinations and having to give up on her plans for a university degree and a Ph.D. There is no telling what she might have accomplished had she not been forced to decide between the narrow options permitted to women at the beginning of the nineteenth century. Had she married Einstein anyway, with a more substantial research background to draw on, Minkowski might not have had to supply to mathematical underpinning for the relativity theory. Or perhaps, if Marić had received her Ph.D., she might not have married Einstein after all—and if Einstein had not been married to a woman with whom he could discuss concepts in physics, and who could edit his papers, would he have produced work as brilliant, or at the same rapid pace? There is no telling. But it is certain that Mileva Marić deserves more remembrance by history than she has heretofore received.

Einstein's Fame Grows

Einstein was somewhat slow to receive the recognition for his four 1905 papers that he deserved. News traveled slowly in the early twentieth century, and more slowly still in the academic world. Einstein had hoped to make something of a splash, to receive recognition from his peers in the scientific community, and possibly also be offered a job at a university—his work at the patent office suited him very well, but it was scarcely the sort of job title that would make other academics take him very seriously. His sister, Maja, reported that he was bitterly disappointed when the publication of his four papers did not meet with the outpouring of critical praise he had hoped for.

However, there were certain people in the field of theoretical physics who recognized what they were looking at once they had Einstein's papers in their hands. The premier physicist of his day was Max Planck, originator of Planck's constant, a concept on which Einstein's relativity theory strongly depended, and he was on the examining board for the *Annelen der Physik*—it had been he who approved Einstein's paper for publication, and Planck himself gave a public lecture on relativity as soon as it had been made public. Planck's approval was precisely the boost that Einstein's reputation needed. He legitimized a highly controversial and revolutionary proposal and made it all the more likely that other physicists would take Einstein's theories seriously. According to physicist Scott Walter,

"A focus on the light-sphere as a heuristic device provides a new perspective on the reception of relativity theory, and on the scientific community's identification of Einstein as the theory's principal architect. Acceptance of relativity theory, according to the best historical accounts, was not a simple function of having read Einstein's paper on the subject. A detailed understanding of the elements that turned Einsteinian relativity into a more viable alternative than its rivals is, however, not yet at hand. Likewise, historians have only recently begun to investigate how scientists came to recognize Einstein as the author of a distinctive approach to relativity, both from the point of view of participant histories…as well as from that of disciplinary history. The latter studies underline the need for careful analysis when evaluating the rise of Einstein's reputation in the scientific community, in that this ascent was accompanied by that of relativity theory itself. Einstein's most influential promoter, Max Planck, himself a founder of relativistic dynamics, was in Einstein's view largely responsible for the attention paid by physicists to relativity theory."

Planck wrote to Einstein, who was only too delighted to correspond with one of the finest scientific minds of his generation. Planck even promised to come to Bern to pay him a visit, and while he ended up having to withdraw from the trip, he sent his assistant, Max Laue, who had also been corresponding with Einstein, in his stead. When Laue arrived in Switzerland, he was astonished to discover that Einstein was employed by the patent office, and not the University of Bern. (Nearly everyone who began corresponding with Einstein after his papers were published were astonished by this discovery.) Laue and Einstein met for the first time in the lobby of the telegraph building, where Einstein's offices were located, but when Einstein first came down to meet him, Laue drew no attention to himself; he was expecting someone older, and he did not think a person as young as Einstein could possible by the author of so many important theoretical papers. They had a long conversation once the mistake was corrected, however, and would go on to be lifelong friends.

The revolutionary nature of Einstein's theories were not the only reason why he was slow to gain the critical following he deserved; anti-Semitism had a role to play as well. Lorentz, a life long opponent of Einstein's theories (partly because his own theories were overthrown by them) wrote that "As remarkable as Einstein's papers are, it still seems to me that something almost unhealthy lies in this unconstruable and impossible to visualize dogma. An Englishman would hardly have given us this theory. It might be here too, as in the case of Cohn, the abstract conceptual character of the Semite expresses itself." Why Jews were thought to be most "abstract" than Gentiles, one can only surmise. But this perception of him undoubtedly played a role in his not receiving any job offers from the academic world. This was still an era of history in which Jews were barely admitted to most traditional European universities as students, however, so the reluctance to take Einstein on as a member of the faculty should come as no surprise.

Even when Einstein humbled himself to apply for position at the University of Bern that involved giving lectures and charging a fee to anyone who attended (attendance being entirely at the discretion of the students), he encountered a faculty that was unaccountably reluctant to hire him, unless undue prejudice played a part. Applicants for that position were supposed to write a short thesis to be considered by the hiring committee, unless they had already accomplished something significant in their field; accordingly, Einstein declined to write a new thesis and instead sent along copies of the seventeen different papers he had published since leaving school in 1900. Only one person on the hiring committee felt that he had done enough to be hired without having to write the thesis; unsurprisingly, Einstein declined to write it. At least the fact that Einstein had finally received his doctorate at last meant that he was promoted at the patent office and received a one thousand francs per year raise. He was still making rather more money that he would have in an academic position.

Chapter Five: Einstein In Zurich

Gravity

In addition to the legend about Einstein failing algebra as a twelve year old, other myths and rumors about his life persist, including one that claims that he gained the inspiration for his theory of gravity by watching a painter fall off a tall ladder and plummet towards the earth. The truth is rather more prosaic than this; Einstein relates that he was merely sitting in his study one day when it struck him that a person who is in free fall is not conscious of their own weight. Rather amusingly, he later said that this was among "the happiest thoughts in my life."

Walter Isaacson describes Einstein's gravitational theory thus:

"Einstein refined his thought experiment so that the falling man was in an enclosed chamber, such as an elevator in free fall above the earth. In this falling chamber (at least until it crashed), the man would feel weightless. Any objects he emptied from his pocket and let loose would float alongside him. Looking at it another way, Einstein imagined a man in an enclosed chamber floating in deep space "far removed from stars and other appreciable masses." He would experience the same perceptions of weightlessness. 'Gravitation naturally does not exist for this observer. He must fasten himself with strings to the floor, otherwise the slightest impact against the floor will cause him to rise slowly towards the ceiling.'

"Then Einstein imagined that a rope was hooked onto the roof of the chamber and pulled up with a constant force. 'The chamber together with the observer then begin to move 'upwards' with a uniformly accelerated motion.' The man inside will feel himself pressed to the floor. 'He is then standing in the chest in exactly the same way as anyone stands in a room of a house on our earth.' If he pulls something from his pocket and lets go, it will fall to the floor 'with an accelerated relative motion' that is the same no matter the weight of the object—just as Galileo discovered to be the case for gravity.

"'The man in the chamber will thus come to the conclusion that he and the chest are in a gravitational field. Of course he will be puzzled for a moment as to why the chest does not fall in this gravitational field. Just then, however, he discovers the hook in the middle of the lid of the chest and the rope which is attached to it, and he consequently comes to the conclusion that the chamber is suspended at rest in the gravitational field.'

"'Ought we to smile at the man and say that he errs in his conclusion?' Einstein asked. Just as with special relativity, there was no right or wrong perception. 'We must rather admit that his mode of grasping the situation violates neither reason nor known mechanical laws.'"

Apparently, prior to Einstein, gravity was a phenomenon thought to hold no more secrets; its properties seemed self-evident.

Academia at Last

By 1908, Einstein had virtually given up on his hopes to be a university professor and establish a career in academia. He had applied to dozens of universities for positions ranging from assistantships to the lowest sort of visiting lecturer to no avail. However, he was still more interested in teaching than in his comfortable, well-paid, interesting work at the patent office, and he had begun seeking work as a teacher of high school science. It wasn't that Einstein was passionately interested in molding the minds of youngsters; but he was having a hard time keeping abreast of all the latest publications in physics because, not being attached to a university, he had no library access. The public libraries were open only when he was at work. He felt that his inability to read the latest publications was having a detrimental affect on the papers he was writing, and more than once he felt the need to beg an editor's pardon in case he was retreating old ground. The schedule of a high school teacher would at least permit him to visit a library within normal business hours, and possibly even have a library of its own.

However, he wasn't certain that he had any qualifications for teaching high school. It had been a number of years since his tutoring jobs and his position as a substitute teacher at a private school. He doubted whether a high school would be especially impressed by the number of scholarly papers he had published, and he felt that his Swabian connections and Jewish appearance would count against him just as much as it had done in the universities.

And perhaps it did, because for a long time he had no luck at all. Finally, he decided to return to the hiring committee at the University of Bern and write the thesis that they so inexplicably required of him in addition to his growing stack of publications. Having done this, he was hired. This resulted in giving him a certain degree of valuable access to the academic world, but as the position paid scarcely anything, he continued to work full time at the patent office. Only three people attended his first lecture; all of them were co-workers from the patent office. Not until the winter term did he attract his first actual student. (Amusingly, no sooner had he got his job at the university than he began allowing his hair to grow into its famous bushy, uncombed halo, which is today so synonymous, not only with Einstein, but with absent-minded professors in general.)

Eventually, even the one faithful student who attended Einstein's winter and spring lectures gave up, and Einstein canceled his lecture series. However, his own doctoral thesis advisor, Alfred Kleiner of the University of Zurich, had recently persuaded his university to create a permanent assistant professorship in theoretical physics, and while initially he meant that it should go to his assistant, a friend of Einstein's named Friedrich Adler, Adler was far more interested in political philosophy than in physics, despite his qualifications, and he suggested to Kleiner that Einstein was a natural fit for the job. Adler wrote to his father of Einstein that he had no talent for charming people into hiring him, being contemptuous of authority and appearances, but that "They [the university authorities] have a bad conscience over how they treated him earlier. The scandal is being felt not only here but in Germany that such a man would have to sit in the patent office."

Adler made it plain to the faculty of the University of Zurich that he was stepping aside in favor of Einstein, and that, in his opinion, it would be the height of foolishness not to hire him. His belief in Einstein's genius was unwavering; being a person of deeply held liberal political beliefs, he may also have suspected the role that anti-Semitism played in spoiling Einstein's efforts to gain academic employment, and thus felt that by lending his support to Einstein's cause, he could help overcome that prejudice. However, Adler's support alone was not enough to secure the post for Einstein. Kleiner turned up at one of Einstein's sparsely attended physics lectures to evaluate him for the professorship, and he made the determination that Einstein was not suited for the job because he was not very good at teaching.

Einstein himself owned that this was the truth; he had never been particularly interested in teaching students, and only wished to gain academic employment so that he would have greater access to research opportunities. Einstein was angry at first, believing that Kleiner had spoiled his chances of ever being considered for an academic post again by repeating his unfavorable assessment of Einstein's teaching abilities to his colleagues. But as it turned out, Kleiner was prepared to give Einstein another chance. He asked him to give another lecture, this time to the university physics society, on a set date that would give Einstein ample opportunity to prepare an impressive presentation. Despite the fact that Einstein resented the entire process of evaluation, he put a great deal of effort into the lecture, and felt afterwards that he had done much better than usual when presented with an audience.

On the strength of this second performance, Kleiner told the university faculty that he believed Einstein should be offered the job; not because his teaching style had improved so very much, but because he would undoubtedly be open to suggestions for improvement, and in any case, he had become too important a figure in the field of theoretical physics not to be given a place in the academic world. Einstein's Jewish heritage was still a mark against him, but Kleiner, apparently, assured his colleagues that Einstein was not visibly Jewish enough to offend their sensibilities. Jews, in the minds of Europeans, were prone to "unpleasant peculiarities", but either because Einstein was secular and had married a Gentile, or because of some other reason, Kleiner believed that Einstein did not exhibit whatever these eccentric characteristics were supposed to be. In the end, the hiring committee made a sort of official ruling on the problem of Einstein's Jewishness— writing an opinion that stands as a testament to the nature of anti-Semitic prejudice in Europe in the early twentieth century:

"The expressions of our colleague Kleiner, based on several years of personal contact, were all the more valuable for the committee as well as for the faculty as a whole, since Herr Dr. Einstein is an Israelite and since precisely to the Israelites among scholars are inscribed (in numerous cases not entirely without cause) all kinds of unpleasant peculiarities of character, such as intrusiveness, impudence, and a shopkeeper's mentality in the perception of their academic position. It should be said, however, that also among the Israelites there exist men who do not exhibit a trace of these disagreeable qualities and that it is not proper, therefore, to disqualify a man only because he happens to be a Jew. Indeed, one occasionally finds people also among non-Jewish scholars who in regard to a commercial perception and utilization of their academic profession develop qualities that are usually considered as specifically Jewish. Therefore, neither the committee nor the faculty as a whole considered it compatible with its dignity to adopt anti-Semitism as a matter of policy."

The faculty's opinion in this case was hardly gracious. It can be summarized as "some Jews possess qualities that we do not like, and for this reason we are suspicious of all Jews; however, we are reasonable enough to admit that not all Jews possess these qualities and that some non-Jews do possess, so in this case we will make an exception, and congratulate ourselves on our open-mindedness." But it was the most generous offer he was likely to get, as he well knew by this point. He declined the initial offer of an appointment because the salary was too low in comparison to what he was being paid by the patent office, but when the committee countered with a higher salary, he accepted, and was finally inducted into the ranks of academia.

Domestic Discord

Einstein's marriage to Mileva Marić was not destined to end happily. They would eventually divorce, and it was around the time of his appointment to the University of Zurich that their relationship began to take a turn into enmity and discord. Einstein had a number of affairs with women throughout his life—so many, in fact, that when he was older, and world famous, a number of young women approached him claiming to be his illegitimate daughter from extra-marital affairs, and though none of their claims were ever substantiated, Einstein was never able to dismiss them out of hand. The first signs of his propensity to stray were showing in 1908, when a young woman whom he had met while staying with his family in Italy wrote him a congratulatory letter after his appointment to the university was mentioned in the newspapers. Einstein had flirted with this young woman during their earlier acquaintance, and when he wrote back to her congratulatory letter, he took up something of the same tone.

However, Marić discovered the young woman's letter, and an ugly confrontation ensued. Einstein began to characterize Marić as being prone to pathological jealousy, apparently unaware that it was perfectly normal behavior for wives to resent their husbands writing flirtatious letters to other women. Marić was beginning to feel that a distance was growing between them, in any case. Now that Einstein was beginning to be famous, and had obtained his university position, gone were the days when his scientific friends would come round to their apartment and include her in their academic discussions on topics related to research in physics. She was relegated more and more to the boring, tedious role of a middle class housewife. Considering how disappointed she was over the unglamorous ending of her own academic career, it is hardly surprising that she was resentful at being excluded from that world of ideas now. So long as Einstein was working in the patent office and relegating his scientific research to the home, she could feel that she was part of that work, and that she still had something to contribute. Now there was nothing for her but to see the world pay more and more attention to her husband's accomplishments, while she cleaned the house and cooked dinner and looked after their son.

Though Einstein had a romantic and flirtatious side that would lead him to seek out relationships with women throughout his life, and though he was apparently very happy in his second marriage, years after his divorce from Marić, in some ways he not naturally suited to domestic life at all. He had always felt that things which were "universal" were more important than those that were "merely personal". This was apparent even during his courtship with Marić, when their letters to one another abstracted their own affection for each other into general principles about the function of the universe. Marić was not unlike Einstein in this; her letters were just as likely as his to dwell more upon articles she had read and lectures she had attended than on plans for their future together. And again, so long as Einstein was excluded from the world of academic research proper, he and Marić could be just as abstract as they liked—but their marriage worked because they did it together. Disdain for the personal is all very well and good when one's personal life is peaceful and easy, but as Marić had discovered, it was not so easy to remain abstract and high minded when there is nothing in one's personal life but boring domestic chores and increasing coldness and misunderstanding with one's partner. Einstein seems to not have been very understanding of this reality; perhaps he never thought the subject worthy of his attention, or else his fierce intelligence might have yielded him some insight on the matter.

The World Begins to Take Notice: Back to Zurich

In 1909, Einstein was offered an honorary doctorate by a German university. He was also invited to give a lecture before an international physics conference in Salzburg, which had reorganized its platform to make a place for his new theories about relativity and the way light travels as quanta, or packets, known today as photons. At this conference, Einstein finally came face to face with the other leading lights in physics of his age, including Max Planck, with whom he had been corresponding for so long, and whose theory of Planck's constant was so integral to relativity. He was expected to lecture on either relativity or light quanta; instead, Einstein chose to discuss some new ideas that he had been formulating on wave theory and molecular theory as it applied to light. The wave – particle duality theory of light would turn out to be just as revolutionary as the rest of his offerings to the field of theoretical physics.

After the conference, it was time for the Einstein household to leave Bern, where they had lived for most of the decade, and return to Zurich, where Marić and Einstein had met as students at the polytechnic school, so that Einstein could take up his position at the university. They were delighted, Marić especially, to be back in this old, familiar city. Their mutual happiness led to a period of reconciliation and harmony in their relationship, and perhaps as a direct result of this, Marić soon discovered that she was expecting their third child. They were thrilled by the pregnancy, and perhaps equally thrilled to discover that their flat was located in the same building as the rooms occupied by Friedrich Adler and his wife, the man who had championed Einstein's cause so effectively to the University of Zurich faculty. Adler and Einstein formed a two-man reading society, not unlike Einstein's Olympian Academy in Bern, retreating to the attic of their building to drink coffee and smoke and read books away from their wives and children. Einstein was baffled by Adler's interest in politics, which had begun to supplant the role of science and physics in his life. It wasn't that he disagreed with the platform of the social democrats, the most liberal of the German parties; it was that Einstein was too suspicious of authority and disenchanted with power structures in general to be loyal to the views of any part. Working within the system to effect change was an alien idea to him. Adler would be much better off, in his view, by returning to physics full time.

Einstein was also given the opportunity to rekindle friendships with some of his former professors and fellow students from his days at school. Marcel Grossman, whose notes had enabled him to pass his mathematics examinations by the skin of his teeth, was still living in Zurich, as was Adolf Hurwitz, a mathematics professor Einstein had offended by skipping most of his classes and lectures. Owing, perhaps, to the fame Einstein had achieved, Hurwitz seemed disposed to forgive the adult man the slights he had been guilty of as a teenager, and the two became friendly. Einstein became a regular attendee of musical evening parties at Hurwitz's house, where he played violin.

When Einstein began teaching at the university, it became evident that he had taken to heart Kleiner's faith in his ability to improve his lecture methods. His students found him unconventional, but interesting—far more so than any of their other professors. Rather than delivering a monologue, as had been his wont when he was lecturing for pay in Bern, he paused at intervals to ask the students if they were keeping up or if they had any questions—a trait which anyone who has ever been a university student knows how to value in a professor—and he didn't mind if his students politely interrupted to make comments or ask questions. This may not sound especially unconventional in this day and age of casual academic styles, but most university professors did not permit this level of classroom interaction with their students. At certain points, Einstein dispensed with lecturing entirely and simply sat down with his students to have conversations. After lectures, it became his habit to invite his students to join him at a nearby café to drink coffee and continue their investigation into the world of theoretical physics. He even invited his more gifted students to become part of his correspondence with Max Planck, when Planck wrote to him with a proof that seemed to contain an error he could not immediately identify.

Eduard Einstein

In July of 1910, a second son was born to Albert Einstein and Mileva Marić, whom they named Eduard but usually called Tete. All of Marić's childbirths had taken a heavy toll on her health, owing to her respiratory problems and hip dysplasia, and this one was especially bad. But rather than allowing Einstein to take on more work so that they could afford the services of a maid, she wrote to her mother to come and join them for a short time.

Einstein faced some criticism for his parenting skills as his sons grew to be adults, particularly in Eduard's case. The distance between father and sons was no doubt partly due to the fact that

Hans Albert and Eduard went to live with Marić after she and Einstein divorced in 1919; the family was living in Berlin at the time, and Einstein remained there after the divorce, while Marić moved back to Zurich and took the children with her. As a young man, Eduard studied medicine with the intent of becoming a doctor, but his academic career, and everything else in his life, was derailed when he suffered a schizophrenic breakdown at the age of twenty. Whether Eduard Einstein's schizophrenia was particularly severe, or whether his condition was complicated by the clumsy and ineffective medical treatments available at the time, is a matter for debate, but by his mid twenties it was apparent that he could not live on his own, and he was committed to an institution, where he remained for the rest of his life. Marić lived nearby and visited him frequently, but Einstein, who moved to the United States in 1933 when the Nazi party began rising to power in Germany, never saw him again.

During his sons' early childhood, however, he was apparently a kind and affectionate father. His older son, Hans Albert, recalled that his father used to build them marvelous toys. He played the violin to amuse them and was happy to look after them when Marić was busy. Perhaps his greatest strength as the father of small children was his powers of concentration; rather than growing irritable when they cried or played at his feet, he was able to continue with his work. Many a parent would grow impatient and snap at their children in such cases; whatever Einstein's other failings as a parent, he seems not to have been guilty of this.

Chapter Six: Einstein in Prague

A New Position

Considering the fact that Einstein was not able to get a position at the University of Zurich until he had gone to extraordinary efforts to improve his teaching abilities, it must have been especially gratifying to him that, when word got around that the University of Prague had offered him a professorship with a pay raise of one thousand francs per annum, his students in Zurich immediately wrote up a petition with a large number of signatures, urging the faculty in Zurich to do whatever they must to retain him as part of the university. According to the student who led the petition, Hans Tanner, "Professor Einstein has an amazing talent for presenting the most difficult problems of theoretical physics so clearly and so comprehensibly that it is a great delight for us to follow his lectures, and he is so good at establishing a perfect rapport with his audience." The University of Zurich accordingly offered to raise his salary by an additional one thousand francs per year.

But the greater prestige of the University of Prague was sufficient to overcome the distasteful prospect of leaving his friends behind and moving to a strange city, and Einstein was inclined to accept the position. However, entrenched anti-Semitism was yet again to briefly mar his prospects. Prague was part of the Austro-Hungarian empire at this time, a society not unlike the militaristic, traditional Prussian one that Einstein had been so disgusted by as a boy in Munich, and professors at Austrian universities had to be approved by no less a person than the Emperor, or at least his ministers. A number of leading physicists, including Max Planck, wrote to the Minister of Education, supporting Einstein's appointment, and he was the first choice of the university faculty. But the imperial authorities refused to allow the appointment of a Jew. Their objections backfired on them, however, when the physicist hired in his place learned that he had been everyone's second choice—he declared that he wanted nothing to do with a university that did not acknowledge merit and ability over prejudice.

Accordingly, Einstein was allowed to take up the position. However, the fact that he was ethnically Jewish proved not to be as insurmountable as the fact that he was a secular Jew—all imperial appointees had to be a member of *some* religion. Einstein dug in his heels on this point for a time, before grudgingly listing his religion on the application forms as "Mosaic". The word "mosaic" sometimes means a hybrid of different beliefs from different cultures and religions, but in the early twentieth century it was also an old fashioned way of saying "Jewish" ("mosaic law" means "the laws of Moses"). Whether Einstein intended the latter meaning, or whether he was being deliberately coy about his nonbelief, is a matter for speculation. He was also required to adopt Austro-Hungarian citizenship in order to take up the position, which he agreed to do so long as he could also keep his Swiss citizenship. Interestingly, from being stateless for a few years after leaving Zurich, he would go on to be a citizen of at least four different countries during his lifetime: Switzerland, Austria-Hungary, Germany, the United States. (He was also offered citizenship in Israel, but he declined.) Einstein finally went to work for the University of Prague in 1911.

Before going to Prague, however, he first went to Leyden, in the Netherlands, to meet with eminent Dutch physicist Hendrik Lorentz, one of the figures in physics he admired more than any other. They got along extremely well; Einstein professed an admiration for Lorentz bordering on love. A mutual colleague had this to say about Einstein and Lorentz's relationship:

"The best easy chair was carefully pushed in place next to the large work table for his esteemed guest. A cigar was given to him, and then Lorentz quietly began to formulate questions concerning Einstein's theory of the bending of light in a gravitational field . . . As Lorentz spoke on, Einstein began to puff less frequently on his cigar, and he sat more intently in his armchair. And when Lorentz had finished, Einstein bent over the slip of paper on which Lorentz had written mathematical formulas. The cigar was out, and Einstein pensively twisted his finger in a lock of hair over his right ear. Lorentz sat smiling at an Einstein completely lost in meditation, exactly the way that a father looks at a particularly beloved son—full of confidence that the youngster will crack the nut he has given him, but eager to see how. Suddenly, Einstein's head sat up joyfully; he had it. Still a bit of give and take, interrupting one another, a partial disagreement, very quick clarification and a complete mutual understanding, and then both men with beaming eyes skimming over the shining riches of the new theory."

Life in Prague

Prague is a larger, dirtier, more sophisticated, and more traditionally European city than Zurich, and the Einsteins did not feel particularly at home there. Marić in particular felt alienated by what she saw as the snobbishness of the people there. But Einstein was making a good salary, and they were able to afford electric lighting, as well as the services of a maid. Einstein found that his office at the university overlooked a courtyard which, twice a day, filled with people who wandered about listlessly and behaved oddly, first in the morning with women, then in the afternoon with men. When he made inquiries, he discovered that the courtyard was adjacent to a lunatic asylum, and the interchangeability of the asylum and university occupants was a subject of great amusement to them.

Socially, Einstein's fame, musical abilities, breadth of reading, and position in society won him a place into "a salon for Prague's Jewish intelligentsia", hosted by socialite Bertha Fanta; as a result, Einstein was able to form a circle of close friends. Throughout his life, people sometimes perceived him as a loner, because he was an iconoclast and an individualist, and because social awkwardness fit the stereotype of a distractedly brilliant professor. But even though Einstein had little patience for small talk or the company of people with whom he shared no intellectual points of reference, he seemed to manage to find a tightly knit circle of likeminded friends wherever he went. Perhaps, as he grew more famous, it was easier for people to perceive him as asocial than to reflect that, with so many people clamoring for his time and attention, he had little time or patience left over for his admirers. He was particularly prone to forming social bonds with other Jews, partly because, especially in academic circles, he could empathize with the artificial obstructions that had been placed in the way of their careers due to anti-Semitism.

Einstein and Marie Curie

In 1911, Einstein was invited to Brussels, in Belgium, to a conference sponsored by a wealthy patron who dabbled in chemistry. The other guests constituted a who's-who list of the most famous minds in various scientific fields, and Einstein interpreted the gathering as more of a vanity exercise than a serious foray into scientific investigation. This first meeting, however, spawned an entire conference series, called the

Solvay Conferences, that would last for two decades.

For Einstein, who seemingly put little thought into his own presentation, the most arresting part of the conference was social rather than scientific in nature. One of the other famous guests was Marie Curie, the Polish born physicist and chemist whose research on radioactivity had garnered her world-wide fame. She and her husband had won the Nobel Prize in 1903 for their work on radiation, and in 1911, during the first Solvay Conference, Curie won the Nobel Prize in chemistry for her discovery of radium and polonium. However, just as her selection was announced, her lover's estranged and abusive wife stole personal correspondence from Curie's flat and had it published in the newspapers; the media firestorm this ignited led to the suggestion that Curie not appear in person to accept the prize, which she dismissed with scorn. Einstein, who considered Curie brilliant, and thought little of the personal indiscretions she was being accused of, wrote her an encouraging letter around this time:

"Do not laugh at me for writing you without having anything sensible to say. But I am so enraged by the base manner in which the public is presently daring to concern itself with you that I absolutely must give vent to this feeling. I am impelled to tell you how much I have come to admire your intellect, your drive, and your honesty, and that I consider myself lucky to have made your personal acquaintance in Brussels. Anyone who does not number among these reptiles is certainly happy, now as before, that we have such personages among us as you, and Langevin too, real people with whom one feels privileged to be in contact. If the rabble continues to occupy itself with you, then simply don't read that hogwash, but rather leave it to the reptile for whom it has been fabricated."

It was little more than a banal interlude in a forgettable conference, but it goes to illustrate Einstein's writing style, his sense of humor, and more importantly, his sense of fairness and justice towards one of the few women in his profession.

Elsa

Mileva Marić was not enjoying the same cultured, varied lifestyle in Prague as Einstein, and when he traveled around Europe, meeting famous physicists and attending conferences, Marić was left behind with the children. It is apparent from her letters to Einstein of this period that she was feeling left out of the loop when it came to his career; she missed having conversations about physics now that Einstein was increasingly absent from the home, and with his friends. A colleague of Einstein's who encountered Marić in Prague while Einstein was away wrote to him that he believed Marić displayed signs of schizophrenia. Apparently, there was some history of that illness in her family, because Einstein wrote back that he was probably right, and that it was "traceable to a…genetic disposition in her mother's family." Considering that their son Eduard would develop severe schizophrenia later in his life, it seems that Einstein was not mistaken about this genetic disposition, but it is difficult to say whether he was being fair in attributing signs of the illness to his wife. Marić undoubtedly had many legitimate reasons for depression and gloom without having to resort to a diagnosis of severe mental illness for an explanation.

In 1912, as his marriage to Marić was growing increasingly unstable, Einstein went to Berlin, where he became re-acquainted with his first cousin, Elsa Einstein; her mother was Einstein's mother sister, and her father was Einstein's father's cousin, Rupert. Elsa was three years older than Einstein, like Marić. Einstein's mother had gone to live with Elsa's parents after the death of her husband. Einstein was thirty three and Elsa thirty six on the occasion of this meeting; Elsa had already been married and divorced, and she had two daughters to raise, Margot and Ilse. She and Einstein had not met since they were young children, though Elsa had many fond memories of listening to Einstein play the violin as a boy.

Einstein seemed to fall in love with Elsa quickly, notwithstanding a brief flirtation with her much younger sister. Elsa was, in most every respect, Marić's complete opposite: matronly, sweet, relatively uncomplicated, not particularly intellectual or academically inclined. A relationship with such a person would have been beneath Einstein's lofty tastes a decade ago, as he proved when he ended his relationship with Marie Winteler. But marriage to Marić had apparently altered his requirements in a romantic partner; now he could think of nothing better than an uncomplicated and steady affection from a woman who did not aspire to be his intellectual equal. Elsa must have fallen for Einstein equally quickly, because they were writing letters to each other as soon as Einstein returned to Prague. Apparently, Einstein had learned his lesson after the last time he exchanged flirtatious letters with a woman, or perhaps the recent indignity that Marie Curie had suffered in having her correspondence stolen was weighing on his mind, because he asked Elsa to send her letters to his office. She was pleased to do so, though she did request that Einstein destroy them after reading.

"I have to have someone to love, otherwise life is miserable," he told Elsa in one of his letters. "And this someone is you."

Partly because of his developing romance with Elsa Einstein, and partly because Germany was the center of academic physics in Europe, Einstein was considering a move to Berlin. The climate in Berlin was, somewhat ironically, more favorable towards Jews than the climate in Bern. In a few years, the universities of Berlin would adopt a doctrine of freely admitting Jewish students, which would later fuel Nazi propaganda that Berlin was "decadent", full of communists and "Jewish intellectuals" that must be purified by Nazi purges. Before the Nazis rose to power, however, Berlin enjoyed a heyday as one of the most stimulating intellectual cities in Europe.

In the mean time, however, the dissolution of Einstein's marriage was becoming so obvious that their older son, Hans Albert, who was about eight at the time, would later say that he had clear memories of the tension growing between his parents. Einstein may have been sensible of the effect that the tension was having on the children, because he apparently made a cursory attempt to end his emotional affair with Elsa.

The fact was, by the end of 1912, everyone in the Einstein family was unhappy. Marić felt isolated and disaffected, Hans Albert was withdrawn and quiet, Eduard was beginning to suffer health problems, which Marić blamed on the poor quality of the Prague city water and the soot polluting the air, and Einstein was growing disgusted with the airs and manners of Austrians. As Einstein had decided there was little chance of being offered a university post in Berlin, and that he had better not move nearer Elsa in any case, he and Marić decided to return to Zurich, where they had first met and fallen in love, and where, for a short time, they had been happy.

Zurich Again

Historians and scientists today who write about the Zurich Polytechnic institute where Einstein and Marić studied refer to it as the ETH, short for the Eidgenössische Technische Hochschule, or, in English, the Swiss Federal Institute of Technology. This is because, in 1911, the school was upgraded to full university status, meaning that it was now permitted to grant doctoral degrees.

Einstein was interested in returning to Zurich, and he was even more interested in teaching at his alma mater. There had been some talk of offering him a position—the faculty at the ETH had asked Einstein to notify them if he received an offer from another institution, and because of this, he turned down an offer from a German university. But when Einstein announced his intention to accept an appointment from the ETH, some of the faculty objected on the grounds that there was no room for a full time professor of theoretical physics at an engineering institute. Furthermore, the old rumors about Einstein being a terrible teacher had caught up to them. A friend of Einstein's in Zurich, Heinrich Zangger, wrote a recommendation for Einstein, addressing these concerns.

"He is not a good teacher for mentally lazy gentlemen who merely want to fill a notebook and then learn it by heart for an exam; he is not a smooth talker, but anyone wishing to learn honestly how to develop his ideas in physics in an honest way, from deep within, and how to examine all premises carefully and see the pitfalls and the problems in his reflections, will find Einstein a first-class teacher, because all of this is expressed in his lectures, which force the audience to think along."

Einstein was touched by Zangger's report, but told him not to bother. If the Zurich faculty was still planning to jerk him around, they could, literally, kiss his ass. (This is a direct translation of the original German in Einstein's letter.)

However, Einstein often pretended, as a matter of pride, to not be interested in positions that he really wanted, once there seemed to be a chance that he wouldn't get them. After a period of frosty demurral, he usually redoubled his efforts to obtain them, and such was the case with the ETH in 1912. Einstein wrote to his prospective employers to assure them that he gave preference to their offer over the offer he had received from Zurich. And a number of people wrote to the ETH in addition to Zanggler to express their support. One of these people was Marie Curie. The eminent Polish-French physicist was motivated both by a genuine esteem for Einstein's genius, and possibly also by a desire to return a favor, considering that he had written to her expressing his support when she received her second Nobel prize amidst a storm of negative publicity. "In Brussels, where I attended a scientific conference in which Mr. Einstein also participated, I was able to admire the clarity of his intellect, the breadth of his information, and the profundity of his knowledge," she wrote to the Zurich faculty. Other physicists who wrote on Einstein's behalf included Henri Poincaré.

The letter writing campaign was successful, and Einstein returned to Zurich with Marić and their sons in 1912. Every member of the family was delighted by the prospect, particularly Marić, who felt that a return to her beloved city would result in new peace of mind. The change of scenery did not have the hoped for effect, however. Marić's depression and gradual withdrawal from day to day life was so evident that it became a matter of public knowledge and concern. Not only was her depression deepening, but her hip dysplasia had been joined by severe rheumatic arthritis, and she was often in severe pain. Einstein's former math professor, Adolf Hurwitz, who had so enlivened their previous residence in Zurich, was still hosting musical gatherings; on one occasion, he made a point of orchestrating an entire evening of performances of compositions by Schumann, Marić's favorite composer, in the hopes of tempting her from home for an evening. Marić attended the concert, but seemed unable to get much enjoyment from it through the haze of various afflictions.

Einstein had put an end to his correspondence with his cousin Elsa some months before moving to Zurich, but as if unable to sever ties completely he had provided her with the address of his new office there. On his thirty fourth birthday she wrote him a short letter, and soon Einstein was suggesting that find a way to make a visit to Zurich for a few days so that they could meet and take a walk together. Einstein had something of a tendency to make great demands of the women in his life without going to a corresponding amount of trouble on their behalf. When Marić became pregnant before their marriage, he seemed content to let her shrift for herself in Zurich while he wrote to her from Italy, enjoying some light flirtation on the side. Likewise, he did not hesitate to invite the divorced Elsa into corresponding with him, despite the fact that a lot more trouble would come to her than to him if they were discovered. Now he was actually proposing that she leave her children behind and travel to Switzerland at her own expense for an illicit rendezvous. In some ways, it seems only good sense on Einstein's part that he had become attracted to qualities such as patience and unconditional loyalty, since any partner of his was bound to need them.

Berlin At Last

Einstein had more or less given up on the prospect of Berlin when he was visited by a group of eminent physicists, including Max Planck, who had traveled to Zurich to offer him an extremely prestigious post: if he accepted, and they succeeded in getting him elected, he would become the youngest ever member of the Prussian Academy of the Sciences, would become the head of a new institution devoted to theoretical physics, and he would become a full professor of the University of Berlin. In addition to all of this glory, the position came with two additional perks: one, his pay would be extremely generous, far more so than his salary at the ETH, and two, he wouldn't have to teach anymore. Instead, he would be a kind of administrator.

Marić was, understandably, less than enthused by the prospect of moving to Berlin. Jews were beginning to be more acceptable in German society, but Serbians like Marić were not enjoying the same social advances. But she resigned herself to doing what was necessary to advance her husband's career. Meanwhile, Einstein was writing to Elsa, excitedly relating to her the news that they would soon be living in the same city. For all practical purposes, Einstein seemed to regard their relationship as permanent and established, but he repeatedly told Elsa that he could not hurt Marić by seeking a divorce—and in any case, even if he had wanted to, in the early twentieth century it was nearly impossible to divorce a spouse unless one possessed evidence of infidelity. Elsa, on the other hand, had her reputation and the futures of her two daughters to think about, and she refused to see their relationship in any terms save as inevitably leading to marriage. She would argue with him about this for years until he finally agreed.

But during their last days in Zurich, Einstein made some attempt to fulfill his duties as a husband and father. He arranged for himself, Marić, and their two sons to join Marie Curie and her daughter Irene on a hiking trip to some of the places where he and Marić had hiked when they were younger and still courting—in other words, he arranged a romantic, but chaperoned, getaway to their old haunts. Eduard, however, became ill just as the trip was beginning, and Marić had to stay at the resort to nurse him. Einstein and Curie and the children went on the trip alone together, during which Einstein took to enthusiastically describing the issues he was wrestling with regarding his gravitational theory. At one point, to the amusement of the children, he grabbed Curie's arm, and, according to her daughter, declared "You understand, what I need to know is exactly what happens to the passengers in an elevator when it falls into emptiness!"

Einstein eventually went on to Berlin alone while Marić remained behind on an extended visit to her parents, during which, without Einstein's knowledge, she had Hans Albert and Eduard baptized in the Serbian Orthodox church. Einstein was merely amused when he discovered this—he had more important things on his mind. There was a brief window of opportunity for him to enjoy Elsa's company alone, before his family joined him in Berlin. Apparently, Elsa cooked for him a good deal, something he much appreciated, despite being famously indifferent to what he ate most of the time.

Chapter Seven: Einstein in Berlin

End of A Marriage

The decline of Einstein and Marić's marriage was swift once the move to Berlin had been effected. Einstein immediately threw himself into long, passionate dialogues with his scientific colleagues, and saw Elsa as much as he could. Marić's spirits withered in Berlin, just as they had in Prague; she ended up having an affair of her own, with a Serbian mathematics professor named Vladimir Varićak, who had challenged Einstein's relativity theory. By July of 1914, four months after their arrival in Berlin that April, the marriage had reached the point of no return; Marić took her children and moved in with her best friend in the city, whose husband also happened to be one of Einstein's closest colleagues.

It seems that Marić did not, by this, intend to provoke a divorce or separation, but possibly to send a message that things had become intolerable and something in their relationship must change. Einstein, however, had lost interest in repairing the relationship. He was prepared to continue living together so that he would not be separated from his children, but he drew up a list of conditions that Marić must agree to before they met again. It can only be surmised what Einstein's intentions were in drawing up this list of conditions, because they are remarkably cold—it seems most likely that despite his claims of wishing to remain together for the children's sake, he was attempting to drive Marić out of his life for good.

The conditions read as follows:

"A. You will make sure

1. That my clothes and laundry are kept in good order;

2. That I will receive my three meals regularly in my room;

3. That my bedroom and study are kept neat, and especially that my desk is left for my use only.

"B. You will renounce all personal relations with me insofar as they are not completely necessary for social reasons. Specifically, you will forego

1. My sitting at home with you;

2. My going out or traveling with you.

"C. You will obey the following points in your relations with me:

1. You will not expect any intimacy from me, nor will you reproach me in any way;

2. You will stop talking to me if I request it;

3. You will leave my bedroom or study immediately without protest if I request it.

D. You will undertake not to belittle me in front of our children, either through words or behavior."

Somewhat incredibly, Marić agreed to these terms initially—but it seems that she still did not fully understand Einstein's intent. He wrote a second time to emphasize that if they were to live together again, it would be as strangers, or friendly business partners. This, it seems, was sufficient to drive the point home. Marić and Einstein began negotiating a separation instead, preparatory to a divorce; she would move back to Zurich, and Einstein would give her half of his salary for the maintenance of herself and the two boys.

The dissolution of his marriage undoubtedly came as a relief to Einstein, who, after all, had a lover waiting in the wings, prepared to marry him as soon as he was free. But he mourned the loss of his sons, and apparently wept bitterly the whole night after they boarded the train for Zurich with their mother. Thereafter, he would only see them a few times a year, on scheduled visits that could not take place in any home that he shared with Elsa.

The one person who was pleased by this outcome of events was Pauline Einstein, his mother, who had hated Marić from the beginning. She had not always got along perfectly well with Elsa, but she regarded her as a much more suitable partner for her son. Elsa's parents were also tolerably pleased. Elsa herself, however, had some grounds for dissatisfaction: Einstein had found the separation so gruelingly painful that he was now gun shy regarding the prospect of a second marriage, and he could not bring himself to marry Elsa immediately. Marić was resistant to granting Einstein a full legal divorce, despite the fact that she had given him grounds for one by leaving him, not to mention having an affair; it is not known whether she had evidence of his affair with Elsa yet, though she certainly suspected it. Einstein did not have the temerity to take Marić to court at this point. He told Elsa that he was afraid that their relationship would lose its fire if they were to marry and settle into domestic normality, but he assured her that "there is no other female creature for me in the world besides you." Elsa was angry and unhappy, but there was little she could do about it.

World War I

Just as Einstein's marriage had collapsed into a bitter war conducted through letters over custody, finances, and disposition of property,

Europe was itself on the brink of world-wide war.

Berlin was a particularly unfortunate place for Einstein to find himself on the eve of the first World War. As a young teenager he had fled Germany and renounced his German citizenship because he was repulsed by the militaristic Prussian culture; now, after living in Switzerland for so many years, he had returned to the heart of Prussian militarism at the very point when so many years of militaristic build up had finally boiled over in the inevitable consequences. Einstein, though he held deep and thoroughly reasoned political beliefs, had never much troubled himself with politicking in the open. He was a socialist and an internationalist, but he had never seen it as his place to advertise those facets of his character. Now, however, was the time to state his pacifist convictions publicly; he saw the war as "irrational" and felt that scientists, as the champions of reason, had a duty to speak out.

Some of Einstein's closest scientific colleagues were adamantly against him in this. Fritz Haber, who had sheltered Marić after she left Einstein, attempted to get an officer's commission in the army, but because he was a Jew, he was forced to join as a non-commissioned officer instead. He turned his scientific talents to producing bombs and poison gases for the German army, much to Einstein's distress. Haber, Max Planck, and other scientific colleagues signed a petition defending Germany's decision to ride roughshod over Belgium's neutrality, and Einstein joined with another Jewish, socialist colleague to create a counter-petition, extolling pacifism. Einstein wrote the text of the petition, reproduced in full below:

"While technology and traffic clearly drive us toward a factual recognition of international relations, and thus toward a common world civilization, it is also true that no war has ever so intensively interrupted the cultural communalism of cooperative work as this present war does. Perhaps we have come to such a salient awareness only on account of the numerous erstwhile common bonds, whose interruption we now sense so painfully.

"Even if this state of affairs should not surprise us, those whose heart is in the least concerned about common world civilization, would have a doubled obligation to fight for the upholding of those principles. Those, however, of whom one should expect such convictions — that is, principally scientists and artists — have thus far almost exclusively uttered statements which would suggest that their desire for the maintenance of these relations has evaporated concurrently with the interruption of relations.

They have spoken with explainable martial spirit — but spoken least of all of peace.

"Such a mood cannot be excused by any national passion; it is unworthy of all that which the world has to date understood by the name of culture. Should this mood achieve a certain universality among the educated, this would be a disaster.

"It would not only be a disaster for civilization, but — and we are firmly convinced of this — a disaster for the national survival of individual states — the very cause for which, ultimately, all this barbarity has been unleashed.

"Through technology the world has become smaller; the states of the large peninsula of Europe appear today as close to each other as the cities of each small Mediterranean peninsula appeared in ancient times. In the needs and experiences of every individual, based on his awareness of manifold of relations, Europe — one could almost say the world — already outlines itself as an element of unity.

"It would consequently be a duty of the educated and well-meaning Europeans to at least make the attempt to prevent Europe — on account of its deficient organization as a whole — from suffering the same tragic fate as ancient Greece once did. Should Europe too gradually exhaust itself and thus perish from fratricidal war?

"The struggle raging today will likely produce no victor; it will leave probably only the vanquished. Therefore, it seems not only good, but rather bitterly necessary that educated men of all nations marshall their influence such that — whatever the still uncertain end of the war may be — the terms of peace shall not become the wellspring of future wars. The evident fact that through this war all European relational conditions slipped into an unstable and plasticized state should rather be used to create an organic European whole. The technological and intellectual conditions for this are extant.

"It need not be deliberated herein by which manner this (new) ordering in Europe is possible. We want merely to emphasize very fundamentally that we are firmly convinced that the time has come where Europe must act as one in order to protect her soil, her inhabitants, and her culture.

"To this end, it seems first of all to be a necessity that all those who have a place in their heart for European culture and civilization, in other words, those who can be called in Goethe's prescient words "good Europeans," come together. For we must not, after all give up the hope that their raised and collective voices — even beneath the din of arms — will not resound unheard, especially, if among these "good Europeans of Tomorrow," we find all those who enjoy esteem and authority among their educated peers.

"But it is necessary that the Europeans first come together, and if — as we hope — enough Europeans in Europe can be found, that it is to say, people to whom Europe is not merely a geographical concept, but rather, a dear affair of the heart, then we shall try to call together such a union of Europeans. Thereupon, such a union shall speak and decide.

"To this end we only want urge and appeal; and if you feel as we do, if you are like mindedly determined to provide the European will the farthest-reaching possible resonance, then we ask you to please send your (supporting) signature to us."

The modern reader can only wonder whether this remarkable document would have had any impact on the political climate in Germany at the beginning of the first World War had it been published; unfortunately, when Einstein was unable to gain the support of Max Planck and other respected colleagues, he decided not to publish it.

The war did more than affront Einstein's political and social beliefs. It cost him a great deal personally, as it became more and more difficult for Germans to travel in Europe. His sons were in Zurich, and while Einstein tried his best to see them, war time restrictions, as well as his ongoing feud with Marić over money made this difficult. He planned one excursion to Zurich around Easter of 1915 that had to be canceled, and when he finally managed to make it to Zurich for three weeks later that year he was only able to see his sons twice. Hans Albert had been writing him for some time, telling him that he had taken up an interest in geometry and science. Einstein both delighted in hearing this and suffered from the knowledge that he could not be present to help his son with his studies. He wrote back to Hans Albert saying,

"I will try to be with you for a month every year so that you will have a father who is close to you and can love you. You can learn a lot of good things from me that no one else can offer you. The things I have gained from so much strenuous work should be of value not only to strangers but especially to my own boys. In the last few days I completed one of the finest papers of my life. When you are older, I will tell you about it."

Despite these domestic problems, and the fact that Europe was falling to pieces around his ears, Einstein was hard at work on expanding his special relativity theory into an expanded general theory of relativity. After months of grueling labor, it was during a series of four lectures to the Prussian Academy in November of 1915 that he unveiled his findings. He was only thirty six years old when he introduced the theory of space-time that overturned the theories of Isaac Newton centuries before and created the model that defines modern physics today. As Walter Isaacson describes it,

"With his special theory of relativity, Einstein had shown that space and time did not have independent existences, but instead formed a fabric of spacetime. Now, with his general version of the theory, this fabric of spacetime became not merely a container for objects and events. Instead, it had its own dynamics that were determined by, and in turn helped to determine, the motion of objects within it—just as the fabric of a trampoline will curve and ripple as a bowling ball and some billiard balls roll across it, and in turn the dynamic curving and rippling of the trampoline fabric will determine the path of the rolling balls and cause the billiard balls to move toward the bowling ball."

Einstein took an enormous amount of personal pride and pleasure in having introduced his discovery to the world. But there was one more hurdle to be got over in his relationship with Marić, and he would need all the comfort that his professional accomplishments could afford him, as dissolving the last bonds of his marriage dominated the next three years of his life.

Divorce

Einstein had initially decided not to pursue a divorce from Marić for a number of reasons, including the fact that he did not wish to distress his sons by visibly supplanting their mother, and because he feared that marrying Elsa would tie him down in an unhappy situation. He could scarcely imagine a marriage that would not end in boredom and tears. But Elsa continued to insist that they must be married, and her parents, who after all were also Einstein's relatives, were even more adamant. So it was that in 1916, Einstein asked Marić to agree to dissolve their marriage permanently. He told Marić that in exchange for a divorce and the right to have his sons visit him occasionally in Berlin, he would increase his financial support from 5600 francs a year to 6000.

Marić wished to see Einstein to discuss the matter personally, and when he refused, she became very ill; Einstein suggested that perhaps she was faking or exaggerating her illness, but even his friends believed that the illness was genuine, induced by emotional stress. Hans Albert Einstein became angry with his father, blaming him for Marić's condition, and stopped writing to him for a time. Though this cold silence did not last, Einstein felt that he had no choice but to abandon the issue of the divorce, at least temporarily.

In 1917, as the war was entering its final stages, Einstein himself became ill with a stomach complaint that would linger in some form for the rest of his life. The doctor attributed it to the poor diet that virtually everyone was restricted to because of war time shortages, and advised him to eat soft, bland meals of bread and plain pasta. His son Eduard, who had been sickly throughout his childhood, was ill as well, which caused Einstein significant distress. He implored his friend Zangger, who was still living in Zurich, to act on his behalf and take Eduard to clinics and sanatoriums. Einstein used the cash prize that he won from the Viennese Academy in 1916 to pay for these treatments.

In 1918, Einstein raised the prospect of a divorce to Marić once again. This time, he offered her 9000 francs a year, 2000 of which would go into a trust fund for his sons; furthermore, he had come to believe that he would, sooner or later, be awarded the Nobel Prize for physics, which would result in a huge amount of money. The full sum of the prize, he told Marić, he would make over to her in full, in exchange for the divorce. Marić had never cared about the money as much as she wished for Einstein to make an effort to reconcile their relationship, but she had to be pragmatic: she, Eduard, and several of her family members were all in poor health, and she needed funds to take care of them. She eventually agreed to the settlement that Einstein proposed, but she refused to relent on the issue of their sons visiting him in Berlin. Einstein was dismayed by this, but chose not to argue with her about it further.

Einstein's marriage to Mileva Marić and the first World War came to an end at roughly the same time, in the winter of 1918. In Zurich, before the magistrates, Einstein admitted to having committed adultery, placing the guilt for the divorce on himself. The divorce decree banned Einstein from marrying again for a period of two years, but this, Einstein was resolved to ignore. He returned to Germany, and six months later, in June of 1919, he and Elsa were finally married.

Albert and Elsa Einstein's marriage was, by all accounts, the model of a stable marital union. They had separate bedrooms, but this seemed to suit them both. Elsa spoke English and French far better than Einstein did; furthermore, she was practical and socially adept in ways he certainly was not. She served as his translator and as the manager of his career as much as the manager of their domestic life. She protected his free time, which enabled him to work diligently on his scientific research. Her two daughters lived with them, Ilsa having been hired as Einstein's secretary and Margot, rather more shy, pursuing her interests in sculpting. Perhaps there was something in Pauline Einstein's advice decades before that her son needed a wife—a subservient, cheerful, affectionate, practical woman—rather than a clever, bookish intellectual partner in order to be happy. Marić, certainly, could never have been satisfied in such a role; it had been her bad luck to fall in love with a man who claimed to want someone with her talents for his life's partner, only to discover that he was not so different from any other man of his time in expecting his wife to be content in a primarily domestic role.

Chapter Eight: Einstein in the World

1919

After World War I ended, Einstein's fame was reaching fabled heights. Europe had been subjected to four years of death, destruction, and nihilism; now the public was eager to put it behind them as "the war to end all wars" and embrace a new positivism. Einstein's general theory of relativity, however little the average person understood it, was nonetheless clearly a major advance in humanity's understanding of the universe, a moral and human triumph after years of utterly senseless conflict. It is hardly surprising that the world embraced him as a hero. Newspaper headlines announcing Einstein's discoveries appeared in London and New York, declaring, "REVOLUTION IN SCIENCE – NEW THEORY OF THE UNIVERSE – NEWTONIAN IDEAS OVERTHROWN". The fact that, according to Einstein, "only 12 men in the world" could understand the book he'd written on relativity seemed only to enhance the world's obsession with him, as if he were a prophet who had received incomprehensible wisdom from a divine being and humanity could only depend on his interpretations to benefit from the blessing. Einstein's work was so much a part of the mainstream in 1919 that Einstein remarked, "now every coachman and waiter argues about whether relativity theory is correct or not."

In response to the public demand for comprehensible explanations of relativity, the New York Times asked Einstein to write an article explaining it. The following long excerpt, from the article, under the subtitle "The Physical Meaning of Geometrical Propositions", provides what is for Einstein a surprisingly layman-friendly explanation:

"In your schooldays most of you who read this book made acquaintance with the noble building of Euclid's geometry, and you remember—perhaps with more respect than love—the magnificent structure, on the lofty staircase of which you were chased about for uncounted hours by conscientious teachers. By reason of your past experience, you would certainly regard every one with disdain who should pronounce even the most out-of-the-way proposition of this science to be untrue. But perhaps this feeling of proud certainty would leave you immediately if someone were to ask you: "What, then, do you mean by the assertion that these propositions are true?" Let us proceed to give this question a little consideration.

"Geometry sets out from certain conceptions such as "plane," "point," and "straight line," with which we are able to associate more or less definite ideas, and from certain simple propositions (axioms) which, in virtue of these ideas, we are inclined to accept as "true." Then, on the basis of a logical process, the justification of which we feel ourselves compelled to admit, all remaining propositions are shown to follow from those axioms, i.e. they are proven. A proposition is then correct ("true") when it has been derived in the recognised manner from the axioms. The question of the "truth" of the individual geometrical propositions is thus reduced to one of the "truth" of the axioms. Now it has long been known that the last question is not only unanswerable by the methods of geometry, but that it is in itself entirely without meaning. We cannot ask whether it is true that only one straight line goes through two points. We can only say that Euclidean geometry deals with things called "straight line," to each of which is ascribed the property of being uniquely determined by two points situated on it. The concept "true" does not tally with the assertions of pure geometry, because by the word "true" we are eventually in the habit of designating always the correspondence with a "real" object; geometry, however, is not concerned with the relation of the ideas involved in it to objects of experience, but only with the logical connection of these ideas among themselves.

"It is not difficult to understand why, in spite of this, we feel constrained to call the propositions of geometry "true." Geometrical ideas correspond to more or less exact objects in nature, and these last are undoubtedly the exclusive cause of the genesis of those ideas. Geometry ought to refrain from such a course, in order to give to its structure the largest possible logical unity. The practice, for example, of seeing in a "distance" two marked positions on a practically rigid body is something which is lodged deeply in our habit of thought. We are accustomed further to regard three points as being situated on a straight line, if their apparent positions can be made to coincide for observation with one eye, under suitable choice of our place of observation.

"If, in pursuance of our habit of thought, we now supplement the propositions of Euclidean geometry by the single proposition that two points on a practically rigid body always correspond to the same distance (line-interval), independently of any changes in position to which we may subject the body, the propositions of Euclidean geometry then resolve themselves into propositions on the possible relative position of practically rigid bodies. Geometry which has been supplemented in this way is then to be treated as a branch of physics. We can now legitimately ask as to the "truth" of geometrical propositions interpreted in this way, since we are justified in asking whether these propositions are satisfied for those real things we have associated with the geometrical ideas. In less exact terms we can express this by saying that by the "truth" of a geometrical proposition in this sense we understand its validity for a construction with ruler and compasses.

"Of course the conviction of the "truth" of geometrical propositions in this sense is founded exclusively on rather incomplete experience. For the present we shall assume the "truth" of the geometrical propositions, then at a later stage (in the general theory of relativity) we shall see that this "truth" is limited, and we shall consider the extent of its limitation."

This short publication of Einstein's became a best-seller in the early 1920's. His fame posed a challenge to him: he both courted the attention given to his work and complained about it to his friends, insisting that he was being hounded to death by reporters and photographers, to the point where he could barely get his work done. But after so many years of struggling for recognition and failing to receive due credit for his work, it is perhaps not so surprising that he threw himself into performing the role of the mad genius for the benefit of the newspapers.

There was one faction of the public that did not fall into Einstein's circle of admirers, however, and that was German nationalists: the people who had started the war, refused to surrender until Germany had been run into the ground, and now bemoaned the decadent, undisciplined condition of moral ruin that German society, in their opinion, was falling into. The abstract nature of Einstein's physics—"Jewish physics", they were quick to point out—contrasted with their notion that order and discipline should extend to all things, from the day to day affairs of domestic life, to the very construction of the universe. Self-promotion amongst scientists was considered a sign that the scientist wasn't very serious about his work, that he didn't care as much about discovery for discovery's sake as he did about making money. And there was a popular view amongst anti-Semites that this was particularly true of Jews, whom they regarded as being overly interested in money and more inclined than Christians to sacrifice scientific principle in the name of profit. And some people, with no particular anti-Semitic motivation, were quite simply resentful of Einstein's fame and success.

In this way, as the twenties wore on, Einstein was acquiring enemies, even as he was becoming famous and universally beloved. As Einstein put it, "By an application of the theory of relativity, today in Germany I am called a German man of science, and in England I am represented as a Swiss Jew. If I come to be regarded as a bête noire, the descriptions will be reversed, and I shall become a Swiss Jew for the Germans and a German man of science for the English!"

But Einstein's loyalties were not for any nation or religion. He was one of the original internationalists, a global citizen before the global community had been thoroughly envisioned. As he explained,

"My passionate sense of social justice and social responsibility has always contrasted oddly with my pronounced lack of need for direct contact with other human beings and communities. I am truly a 'lone traveler' and have never belonged to my country, my home, my friends, or even my immediate family, with my whole heart; in the face of all these ties, I have never lost a sense of distance and a need for solitude."

Between the Wars: Weimar Berlin and After

In post World War I Germany, the political system was made up of a large number of competing political parties, and a significant number of them—the various nationalist parties that eventually coalesced into the National Socialist, or Nazi party—adopted openly anti-Semitic goals. Some German Jews attempted to protect themselves from persecution by assimilating, wiping away most outward signs of their Jewishness. Einstein, on the other hand, became only more open and vocal about being a Jew. Among other things, he began to campaign for the establishment of Hebrew University in Jerusalem; he was also a grudging and somewhat conflicted supporter of establishing a Jewish homeland in Palestine. "People need a scapegoat, and the Jew is responsible," he remarked. He did not believe that he or any of his fellow Jews would gain any protection against anti-Semitism by making an effort to seem less Jewish; to anti-Semites, Jews were despicable because of their inborn qualities, and no amount of assimilation could change blood.

In 1921, Einstein went to America for a tour to raise funds for the Hebrew University. His reception was extraordinary. As Walter Isaacson puts it, "The world had never before seen, and perhaps never will again, such a scientific celebrity superstar, one who also happened to be a gentle icon of humanist values and a living patron saint for Jews." During his American tour, he laid eyes on Princeton for the first time, the university which, in twelve years, would become his final academic home. Later in 1921, he won the Nobel Prize, just as he had predicted.

Fleeing Germany

By the early 1930's, Germany was no longer a safe place for Einstein to live. Just two years earlier, on the occasion of his fiftieth birthday, the city of Berlin had chosen to honor its most famous scientist (and by this time, Germany's most popular and internationally respected citizen) by givig him a weekend cottage and a small plot of land for holidays and short retreats. But even though Hitler was not yet Chancellor of Germany, the political situation was already visibly unstable; President of the Reichstag Paul von Hindenberg was nearing the end of his life, and the chancellor he had selected had little popular support in comparison to Hitler. While even a genius of Einstein's caliber could not fully predict all the changes that the Nazis would bring to Germany and the world over the next twelve years, he was more than politically savvy enough to understand that life in Berlin, not just for Jews, but for anyone who did not fall in line with the Nazi agenda, was about to become impossible.

The 1920's had not been an ideal time in his life; his family life had seen some upsets, owing to Einstein's multiple affairs, his objection to the woman that his now adult son Hans Albert married, and his son Eduard's attempted suicide and institutionalization for schizophrenia. By the dawn of the 1930's, he had ceased to be as competitive for honors in physics as he had been in his youth, but his participation in politics was increasing. Indeed, it would scarcely have been possible for a man as outspoken as Einstein, who possessed deeply cherished principles of social justice, not to become politically outspoken as the Nazi party rose to power. Jokingly, he described himself as "a militant pacifist". He was so well known for his antipathy towards nationalism that he was openly recruited by communist political parties to lend his support to the cause of socialist rebellion. But he had little more patience for the near-religious ideologies espoused by the communist parties than he had for the fascist parties. As Isaacson puts it, Einstein believed that "freedom and individualism are necessary for creativity and imagination to flourish." Doctrinaire belief in any single political party was, to him, antithetical to such intellectual freedom.

He had made a second visit to America in 1930, where he spent most of his time lecturing in southern California. He found America rather banal and boring compared to Zurich and Holland, but when he returned to Berlin in 1931 he forced to reflect that there might be nowhere left besides America where the climate for intellectual expression remained free. By June of 1932, he had received an offer to join the faculty of Princeton University for $15,000 a year. He had been offered similar positions at Oxford and at the California Institute of Technology, and he was still considering whether he could accept more than one of them. But one way or another, an immediate relocation to the American continent was necessary. Einstein set sail for the United States on the steamer ship *Oakland* in December of 1932 with thirty pieces of luggage—a clear sign that he was going to America to stay. His timing could scarcely have been better. One month later, on January 30, 1933, Hitler was named Chancellor of Germany; in April of that year, he would name himself *führer* and abolish the Reichstag, and with it democratic government. Jews in Germany began to feel the consequences virtually overnight. Einstein had not been certain that he was leaving Berlin permanently when he first set out for America, but after the election he told a reporter, "As long as I have any choice in the matter, I shall live only in a country where civil liberty, tolerance and equality of all citizens before the law prevail. These conditions do not exist in Germany at the present time."

Almost immediately after Hitler became Chancellor, there was a Nazi raid on Einstein's Berlin flat; Elsa's daughter Margot received warning just in time to get Einstein's papers sent to Paris. Einstein turned in his passport and renounced his German citizenship for a second time upon learning of this, and resigned his membership in the Prussian Academy. In doing so, he struck a clever blow against the Nazi propaganda machine—they had been planning to formally oust him from the Academy themselves, and denounce him as a corruptive influence.

The brutality of the Nazi regime was so extreme that Einstein came to abandon the pacifistic stance he had adopted during the first World War. He wrote public letters to various pacifist societies he had worked with in the past, indicating that his feelings had changed: Germany's aggression was so extreme that, especially in her neighboring countries, Belgium and France, preparing their armies could only be looked on as an act of self-defense. The same man who, as a teenager, had cast off his German citizenship to avoid being conscripted into the army, made an extraordinary declaration in one of these letters:

"I must tell you candidly: Under today's conditions, if I were a Belgian, I would not refuse military service, but gladly take it upon me in the knowledge of serving European civilization. This does not mean that I am surrendering the principle for which I have stood heretofore. I have no greater hope than that the time may not be far off when refusal of military service will once again be an effective method of serving the cause of human progress."

After moving the majority of his possession to the United States, Einstein returned to Europe for a lecture tour, tracing a circuitous path around Germany as he visited England, Belgium, the Netherlands, and paid one final visit to his son Eduard in Zurich. Einstein and Marić had significantly repaired their relationship over the years since their divorce, and they wrote many amiable letters to each other regarding their sons. On this visit to Switzerland, Marić went so far as to invite Einstein and Elsa to stay with her so that Einstein would be conveniently located for paying visits to the institution where Eduard resided.

At that time, Einstein assumed that he would make a visit to Europe at least once a year, however long he remained resident in the United States, and see Eduard and Marić as often as could be managed. However, unbeknownst to him, when he departed for the United States again on October 7, 1933, it would be for the last time. He was never able to return to Europe again.

Chapter Nine: Einstein in America

Princeton

"[In Princeton] Einstein soon acquired an image, which grew into a near legend but was nevertheless based on reality, of being a kindly and gentle professor, distracted at times but unfailingly sweet, who wandered about lost in thought, helped children with their homework, and rarely combed his hair or wore socks."

Walter Isaacson

Einstein's celebrity continued unabated during his years at Princeton, where he worked from 1933 until his death in 1955. He was a popular figure locally, known for giving impromptu violin performances to strangers and exuding a deep sense of caring about the state of the world and the condition of humanity. In this latter period of his career, matters pertaining to social justice seemed to preoccupy him a great deal more than making new breakthroughs in the field of theoretical physics. Of course, he had made so many paradigm-changing discoveries in physics by the age of thirty-six that if he had abandoned science entirely before the age of forty, he would still be remembered as the greatest genius of the modern scientific age. It is scarcely surprising that he felt that his talents and his fame could be put to better use, especially once he found himself living in an age of terrifying violence and fascist dictatorships.

American society suited Einstein in a lot of ways. He felt that anti-Semitism was virtually nonexistent in the United States—an assessment which seems a trifle naïve, although compared to the virulent anti-Semitism in Europe, it is perhaps understandable that America, a younger and more diverse society, seemed to be something of a haven in that regard. And he believed that, in America, it seemed easier for a person to achieve success based on their merits, compared to in Europe where issues of class, tradition, ethnicity, and wealth seemed to him to have a more defining role in a person's career. Individualism and freedom of speech were American traits that Einstein prized highly. Indeed, both Einstein and Elsa were so happy in America that they sometimes felt guilty, knowing what their Jewish friends and colleagues in Europe were suffering as Nazi aggression intensified. The Einstein household became completely consolidated in Princeton in 1934, when Elsa's daughter Ilsa died, and her daughter Margot moved to New Jersey to be with them. Einstein began proceedings to become an American citizen in 1935, and would complete the process in 1940.

The Death of Elsa Einstein

Elsa Einstein became ill with heart and kidney problems early in 1936. Einstein divided his time between sitting at her bedside to read to her and engrossing himself in his work as a distraction, his lifelong refuge when personal troubles threatened to overwhelm him. In the hopes of speeding her recovery, Einstein took her to a remote cabin in the Adirondack Mountains for the summer, where they had a peaceful vacation together. She seemed to improved for a short time, but in the winter, she began to decline, and ultimately passed away in December of that year.

Elsa's death affected Einstein profoundly. He cried for the first time since his mother's death decades before, and retreated into silence and isolation, declining social invitations and spending long hours at his office. But he was not bereft of human contact; he had a wide circle of friends who made efforts to entice him out of his "cave", as he called it, and his secretary and step-daughter were still living at home with him.

Eventually, his sister Maja, with whom he had always been close, moved in with them as well. Maja had been living in Italy ever since her family moved there when she was a girl, but she was forced to flee when Mussolini came to power and began driving Jews from the country. Einstein was delighted to have her with him. Two years later, Einstein's son, Hans Albert, also came to America with his wife and two sons and accepted a professorship in engineering at Clemson University in South Carolina, and Einstein visited him there frequently, particularly after Hans Albert's six year old son died of diphtheria. Inquiries were made about sending Eduard Einstein to America as well, but because of his mental health diagnosis, the American authorities would not permit him to immigrate. He remained institutionalized until the end of his life, and Mileva Marić attempted to care for him as best she could, even after her sister Zorka died of alcoholism, leaving her essentially alone in the world.

As world politics barreled relentlessly towards the outbreak of the second World War, Einstein maintained his unceasing efforts to raise public awareness about the dangers of fascism and anti-Semitism. In 1938, Einstein wrote a revealing and perceptive article for *Collier's Magazine*, entitled, "Why Do They Hate the Jews?", in which he attempted to explain to American readers the method by which Nazi propaganda had persuaded a nation of ordinary Germans to collude in the persecution of Jews:

"I should like to begin by telling you an ancient fable, with a few minor changes - a fable that will serve to throw into bold relief the mainsprings of political anti-Semitism:

"The shepherd boy said to the horse: "You are the noblest beast that treads the earth. You deserve to live in untroubled bliss; and indeed your happiness would be complete were it not for the treacherous stag. But he practiced from youth to excel you in fleetness of foot. His faster pace allows him to reach the water holes before you do. He and his tribe drink up the water far and wide, while you and your foal are left to thirst. Stay with me! My wisdom and guidance shall deliver you and your kind from a dismal and ignominious state."

"Blinded by envy and hatred of the stag, the horse agreed. He yielded to the shepherd lad's bridle. He lost his freedom and became the shepherd's slave.

"The horse in this fable represents a people, and the shepherd lad a class or clique aspiring to absolute rule over the people; the stag, on the other hand, represents the Jews.

"I can hear you say: "A most unlikely tale! No creature would be as foolish as the horse in your fable." But let us give it a little more thought. The horse had been suffering the pangs of thirst, and his vanity was often pricked when he saw the nimble stag outrunning him. You, who have known no such pain and vexation, may find it difficult to understand that hatred and blindness should have driven the horse to act with such ill-advised, gullible haste. The horse, however, fell an easy victim to temptation because his earlier tribulations had prepared him for such a blunder. For there is much truth in the saying that it is easy to give just and wise counsel - to others! - but hard to act justly and wisely for oneself. I say to you with full conviction: We all have often played the tragic role of the horse and we are in constant danger of yielding to temptation again."

Einstein's campaigns for social justice were not limited to the plight of Jewish refugees from

Europe, thought this cause was obviously \
near to his heart. He pitted himself against
racism in the United States as well. When the
famous black opera singer Marian Anderson was
invited to perform in Princeton, she was refused
lodgings at the best hotel in the city, and Einstein
drew public attention to this injustice by
personally inviting her to stay as a guest in his
own home instead. They became lifelong friends
after this encounter, and Anderson visited him in
Princeton several more times before his death in
1955.

The Atomic Bomb

In the early 1930's, Einstein predicted that Hitler
would regret driving every Jewish scientist out of
Germany, and by extension Europe, if war broke
out. He did not then have any conception of how
right he was, or how the collaboration of refugee
Jewish scientists would lead to the production of

nuclear weaponry. Decades before, he had been approached about the possibility of producing enormous energy by splitting the atom, but he had considered the idea impractical. In 1939, however, an old friend of his, Hungarian physicist Leó Szilárd, approached him with a serious problem: could Einstein, who was close friends with the Queen Mother of Belgium, warn the Belgians that the Germans might attempt to seize shipments of uranium from their colony in Congo? The danger, he explained, lay in the theoretical process by which "an explosive chain reaction could be produced in uranium layered with graphite by the neutrons released from nuclear fission." Einstein was astonished by the suggestion, but he quickly grasped the seriousness of the situation.

Due to his status as a refugee in America, Einstein was advised that it would be better to relay this warning through American channels. At first he thought of approaching the State

Department; but he had visited President Roosevelt personally a few years earlier, and believed that the warning would be more effective if he addressed the letter directly to the president and went to Washington himself to deliver it. The pertinent passage of the letter is quoted below:

"It may become possible to set up a nuclear chain reaction in a large mass of uranium, by which vast amounts of power and large quantities of new radium-like elements would be generated. Now it appears almost certain that this could be achieved in the immediate future."

Einstein was very much aware that his advice would probably lead to the United States building such a bomb, but he believed that even this was preferable to permitting Nazi Germany to beat the rest of the world to it, particularly after Germany shocked the world by invading Poland in 1939. In the end, it was not Einstein who delivered the letter, but a friend of his, Alex Sachs, who, rather than taking the risk that the letter would become lost in the shuffle of presidential paperwork, chose to read it aloud in the Oval Office. Roosevelt took action immediately, drawing up a committee to investigate Germany's attempts at building such a bomb, and this led directly to the organization of the Manhattan Project.

Einstein's letter to Roosevelt was his most significant contribution to the building of the atomic bomb. He played no formal role in the Manhattan Project, though he was consulted by some of the scientists involved on a problem involving the separation of isotopes. He was not interested in being more involved in the project; he was a theoretical, not a nuclear physicist. Furthermore, he probably would not have been allowed to participate even if he had wanted to.

Einstein's universal fame and huge popularity with the public did not protect him from the suspicion of American authorities who had paid close attention to his socialist, internationalist, pacifistic politics. During the second World War, and indeed for the rest of his life, he was suspected of having communist sympathies and passing information to the Soviets, despite the fact that he was just as public about his antipathy to the Russian dictatorship as he was about his opposition to German fascism. Einstein had, in fact, been invited to Russia on a number of occasions, but he had never accepted any of these invitations because he did not wish his doing so to be construed as an expression of support for the Soviet regime.

It was as if, because he was opposed to Soviet communism on the grounds that it suppressed the liberty of the individual and curtailed human rights, rather than on grounds of American nationalism, his loyalty could not be trusted.

Einstein's FBI file concluded that he could not be employed by the American government on any secret projects. It did not, however, stop Einstein from taking a citizenship test and publicly celebrating his formally receiving American citizenship in the summer of 1940.

A friend of Einstein's who was deeply involved in the Manhattan Project paid him a visit when the bomb was close to completion, and Einstein was left deeply unsettled. He intuited immediately that everything about modern warfare and global balance of power politics would be irrevocably altered by the existence of such a weapon. He felt, however, that the politicians and military leaders who would make the decision to deploy such a weapon were not considering this as deeply as they should. Einstein seriously debated the possibility of alerting the general public about the existence of the atom bomb, so as to force the politicians into regarding the implications of using it. Einstein had always believed that the best chance for world peace was international federalization—something similar to the United Nations, but more powerful, with a larger army, subject to multi-national decision making processes. Any single nation having possession of a weapon like the atom bomb could ruin any chance of such international cooperation.

It took the urging of his old friend and fellow physicist Niels Bohr to persuade Einstein not to go public, both for his own safety and the safety of others. Einstein made one last attempt to exert his influence in the matter of atomic warfare, writing a second letter to Roosevelt, pointing out that Germany was close to defeat and clearly had not developed atomic capabilities. Using the American bomb in Japan, he argued, was unnecessarily risky and destructive. What impact this letter might have had on history can only be speculated upon: Roosevelt never read it. The letter was found unopened amongst his papers after his death. His successor, Harry Truman, did read it, but Einstein's advice was unheeded.

After the War

Einstein was devastated when the United States dropped atomic bombs on Hiroshima and Nagasaki. Once the existence of the bombs became public knowledge, Einstein was woven into the narrative of their creation; his first letter to Roosevelt was treated as the starting point in the bomb's development. The perception that Einstein was deeply involved in creating the bomb and that he approved of its use followed him for the rest of his life; it is a myth that persists to this day. Einstein denied this story whenever he had the chance, publicly declaring that, had he known that Germany's attempts to build the bomb would be unsuccessful, he "would never have lifted a finger".

After the war, Einstein found himself reflecting on the example of Alfred Nobel, who had invented dynamite for use in mining, and afterwards felt so guilty about its use in weaponry that he established the Nobel Prize as a kind of atonement for having ushered the source of so much destruction into the world. Einstein felt a similar need to atone his role in the development of the atomic bomb: his project, from the end of the war until the time of his death, was to promote the one-world government, an alliance of nations which would act jointly to wield authority over the use of atomic weapons and which would have the authority to intervene in any nation where a majority was oppressing a minority.

Einstein's efforts had some effect. President Harry Truman made some attempts to organize an international authority that would preside over atomic weaponry, but the era of the Cold

War was fast approaching, and the Soviet Union was already unwilling to cooperate with the United States.

Einstein's Final Years

Einstein retired in 1948, when he was 66. He continued to come into his office, take walks around Princeton, and delight visitors who had chance encounters with the world's most famous scientist. He spend long hours in conversation with Heisenberg, author of the uncertainty principle, and Gödel, author of the incompleteness theory of mathematics; the three of them together had completely dismantled the deterministic scientific models of the nineteenth century and put twentieth century science on a path towards abstraction, uncertainty, and mystery. Einstein, however, was to spend the rest of his life working on a theory of everything, an attempt to interpret the laws of the universe in a way that would make his own relativity theories a little less abstract. He was not to finish this work before his death.

In 1948, Mileva Marić, whose later years were devoted entirely to the care of their younger Eduard, suffered a fall on ice, and lapsed in and out of a coma before dying in May of that year. Eduard was still institutionalized, and Einstein promised to see to it that he was taken care of for as long as he lived, even if he bankrupted himself in the process.

Around the same time that Marić died, Einstein began to grow ill with the same stomach complaint that had left him bedridden in Berlin a few decades before. Surgery revealed an aneurysm in his abdominal aorta. There was no available treatment; the aneurysm would inevitably burst, and kill him when it did so, but he could delay the inevitable for a few years if he ate well and avoided stress. Einstein took this diagnosis in stride. He had suffered the loss of many people whom he loved—his parents, then his daughter. His youngest son was unreachable, his wife Elsa had died, and his sister Maja died shortly after he himself became ill. But he lived his last years in the company of his step-daughter Margot and his long-time secretary Helen Dukas, and he was surrounded by friends, colleagues, and former students who not only revered him for his scientific accomplishments, but loved him deeply as a person.

In 1952, upon the death of Einstein's friend and Israel's first president, Chaim Weizmann, the presidency of Israel was offered to Einstein. Despite the fact that he was never a wholehearted supporter of the Jewish state in Palestine, he was the world's most famous and beloved Jew, and the choice seemed an obvious one to many. They were even prepared to let him continue his scientific work in peace, albeit from

Jerusalem. Einstein was deeply moved by the offer, but he immediately recognized the impossibility of accepting it. He was a physicist; all his life he had pursued objective truths. The duties of a president would require abilities he did not possess, such as the ability to compromise. When the official representative appeared at his home to make the request, Einstein had already written his letter of refusal.

Einstein had become very ill by the time he turned 76 in 1955. When he received word that his old friend from the Zurich Polytechnic school, Michele Besso, had died, he seemed to regard it as a sign of his own imminent mortality. "He has departed from this strange world a little ahead of me," Einstein wrote. "That means nothing. For us believing physicists, the distinction between past, present and future is only a stubborn illusion."

In April of 1955, Einstein's aortic aneurysm began to rupture. He collapsed at home, and was visited by a doctor, who administered morphine to help him with the pain. Anxious to preserve the life of one of the world's most beloved figures, a group of leading physicians assembled in Princeton to discuss the possibility of performing a surgery that might repair the aneurysm, but Einstein was not interested. He felt that his life was coming to its natural conclusion, that it would be "tasteless" to attempt to extend it "artificially". When the pain became too intense to bear, Einstein was rushed to the hospital, and his son Hans Albert flew to Princeton to be with him. He recovered briefly, long enough to spend a final day making notes on his theories and visiting with his family, who remained clustered around him. Then, in the early hours of the morning on April 18, 1955, the aneurysm burst, and Einstein died instantly.

According to instructions given before his death, Einstein was cremated and his ashes scattered so that his burial site would not become a monument for pilgrimage. However, rather curiously, Einstein's brain was removed from his body approximately seven hours after his death by Thomas Harvey, the coroner who performed his autopsy. He did not have permission from Einstein's family to do this. When Hans Albert Einstein heard of this, he called to demand that it be cremated, but Harvey insisted that the brain of such a brilliant man would probably yield useful information if it were studied, and that surely a scientist like Einstein would approve of this. Reluctantly, Hans Albert relented.

Bizarrely, Einstein's brain remained in Harvey's possession until 1998, when he turned the remaining pieces over to Princeton University's pathology lab. Over the years, he gave pieces of it to nearly any person claiming research credentials who asked him for it. Of those who received a piece of Einstein's brain, scarcely any produced publishable research. These findings, or the lack thereof, probably would not have surprised Einstein himself. As he pointed out on multiple occasions throughout his life, he did not feel that he had any special talents. "I am only passionately curious," he explained.

Walter Isaacson relates the details of a conversation that Einstein had with a poet named Saint-John Perse in the 1930s', which may explain more about Einstein's thinking than any mere accident of biology that can be observed in studying the physical remains of his brain:

"'How does the idea of a poem come?'" Einstein asked. The poet spoke of the role played by intuition and imagination. "'It's the same for a man of science,'" Einstein responded with delight. "'It is a sudden illumination, almost a rapture. Later, to be sure, intelligence analyzes and experiments confirm or invalidate the intuition. But initially there is a great forward leap of the imagination.'"

Modern science owes much to Einstein's brain: probably not because it possessed a higher number of glial cells than that of the average human, but because it was the seat of this remarkable imagination.

The End

Other books available by author on Kindle, paperback and audio:

Nikola Tesla: Prophet of the Modern Technological Age

Further Reading

Einstein, His Life and Universe, by Walter Isaacson

Autobiographical Notes, by Albert Einstein

> https://archive.org/stream/EinsteinAuto biography/EinsteinAutobiography_djvu.t xt

What Was Albert Einstein's True Relationship to Judaism—and Zionism?

> http://forward.com/news/325189/what- was-einsteins-relationship-to-judaism- and-zionism/#ixzz47q8L3uq8

Einstein and His Love of Music

> http://www.pha.jhu.edu/einstein/stuff/ei nstein&music.pdf

The Education of Albert Einstein

> https://einsteingame.web.cern.ch/einstei
> ngame/otherswork/Vlado/links/Ein_Sym
> p92.pdf

Albert Einstein's Love Letters

> https://www.brainpickings.org/2015/07/
> 27/albert-einstein-mileva-maric-love-
> letters/

The Year of Albert Einstein

> http://www.smithsonianmag.com/scienc
> e-nature/the-year-of-albert-einstein-
> 75841381/#S0dzIAyOH37Dxlvf.99

Figures of Light in the Early History of
Relativity, by Scott Walter

http://philsci-archive.pitt.edu/9134/4/walter2012-06-02.pdf

Manifesto to the Europeans

http://www.onbeing.org/program/einstein039s-god-einstein039s-ethics/extra/einstein-manifesto-europeans-1914/1987

Relativity: The Special and General Theory

http://www.bartleby.com/173/

Why Do They Hate the Jews?

http://www.filosofiaesoterica.com/ler.php?id=1397#.VzUGsKODGko

42772744R00145

Made in the USA
Middletown, DE
21 April 2017